LIFEScience
Lab Manual

purposeful design®
p u b l i c a t i o n s

Colorado Springs, Colorado

Printed in the United States of America
22 21 20 19 18 17 16   1 2 3 4 5 6 7

Life Science – Lab Manual
Purposeful Design Science series
ISBN 978-1-58331-542-2, Catalog #20073

Purposeful Design Publications is the publishing division of the Association of Christian Schools International (ACSI) and is committed to the ministry of Christian school education, to enable Christian educators and schools worldwide to effectively prepare students for life. As the publisher of textbooks, trade books, and other educational resources within ACSI, Purposeful Design Publications strives to produce biblically sound materials that reflect Christian scholarship and stewardship and that address the identified needs of Christian schools around the world.

References to books, computer software, and other ancillary resources in this series are not endorsements by ACSI. These materials were selected to provide teachers with additional resources appropriate to the concepts being taught and to promote student understanding and enjoyment.

Unless otherwise noted, all Scripture quotations are taken from THE HOLY BIBLE, NEW INTERNATIONAL VERSION®, NIV® Copyright © 1973, 1978, 1984, 2011 by Biblica, Inc.® Used by permission. All rights reserved worldwide.

Purposeful Design Publications
A Division of ACSI
PO Box 65130 • Colorado Springs, CO • 80962-5130
Customer Service Department: 800/367-0798 • Website: www.purposefuldesign.com

Name: _____ Date: _____

Lab 1.1.2A Dead or Alive?

QUESTION: How is it possible to distinguish if yeast is alive or dead?

HYPOTHESIS: _____

_____

EXPERIMENT:

You will need:	• water	• thermometer	• 2 test tubes
• two 3 g yeast samples	• beaker (100 mL)	• marking pencil	• test tube rack
• 2 petri dishes	• graduated cylinder (50 mL)	• spoon	• sugar
• hand lens	• hot plate	• hot mitt	• stirring rod

Steps:

1. Place one yeast sample into a petri dish and the other sample into a separate dish. Set the petri dishes side by side. The petri dish on the left will be Sample A. The petri dish on the right will be Sample B. Examine both samples of yeast with a hand lens. Draw and describe your observations.

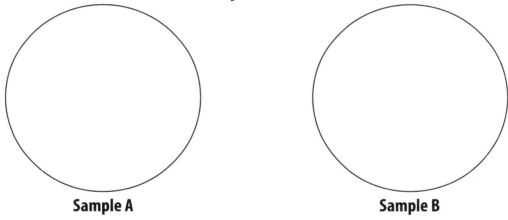

Sample A   **Sample B**

_____   _____

_____   _____

2. Heat 50 mL of water in a beaker on the hot plate until it reaches 30°C.
3. While the water is heating, use the marking pencil to label one test tube *A* and the other *B*. Place test tubes in test tube rack.
4. Hold the beaker with a hot mitt and pour 15 mL of the water into each test tube.
5. Use the spoon to add a small amount of sugar to each test tube. Stir the contents with a stirring rod.
6. Add the yeast sample that was on the left to Test Tube A and the yeast sample on the right to Test Tube B. Stir gently.
7. Record your observations at 5-minute intervals for 20 minutes.

Lab 1.1.2A Dead or Alive?

Yeast Observations

	A	B
5 minutes		
10 minutes		
15 minutes		
20 minutes		

8. Draw your observations after 20 minutes.

A

B

ANALYZE AND CONCLUDE:

1. Compare and contrast the yeast samples before the experiment. _____

_____

2. Compare and contrast the yeast samples after the experiment. _____

_____

3. Was either sample alive? _____ If so, which one? _____

How could you tell? _____

_____

4. Consider the characteristics of living things. Think of another way to determine

whether the yeast was alive or dead. _____

_____

5. Compare your hypothesis to the results. _____

_____

Name: _____ Date: _____

Lab 1.1.4A Healthy Food

QUESTION: Does having a high amount of organic compounds make food healthy?

HYPOTHESIS: _____

_____

EXPERIMENT:

You will need:	• various packaged foods

Steps:

1. Look at the front label of the packaged foods. Record which packaged food you think is the healthiest and explain your decision. _____

_____

_____

2. Which packaged food do you think is the least healthy? Why?

_____

_____

3. Look at the nutrition labels on each food package to determine the amounts of organic compounds in the food. Record the percentages on the chart. If a percentage is not given, record the number of grams.

Organic Compound Percentage in Food

Food Item	Total Fats	Total Carbs.	Dietary Fiber	Sugars	Protein

Lab 1.1.4A Healthy Food

ANALYZE AND CONCLUDE:

1. What were the organic compound percentages of the food that you chose as the healthiest? _____

_____

2. What were the organic compound percentages of the food that you chose as the least healthy? _____

_____

3. Research how to determine if foods are healthy or unhealthy. What three key ingredients should you watch out for? _____

_____

Which of those ingredients is not an organic compound? _____

4. Based on those three ingredients, were your choices for healthiest and least healthy correct? Explain why. _____

_____

5. Which food do you think is the healthiest now? _____

Which do you think is the least healthy? _____

6. Why is it important to eat foods with organic compounds? _____

_____

7. Do you think that people need the same amount of every organic compound? Explain. _____

8. Does having a lot of one organic compound mean that a food is healthy? Why?

_____

_____

9. Why is eating a wide variety of foods important? _____

_____

10. Compare your hypothesis to what you have learned from this investigation.

_____

_____

Needs of a Living Thing

Choose an organism and observe it in its natural habitat.

Common name of organism _____

Scientific name of organism (genus and species) _____

Native countries _____

_____

Habitat _____

_____

Food _____

_____

Water _____

_____

Gases _____

_____

How does it use energy? _____

_____

_____

How much space does it need? _____

_____

_____

Organic Compounds

Imagine that your friends come to you for advice on how to improve their diets. Given your knowledge of organic compounds, what would you tell each of them?

1. Elizabeth has decided to become a vegetarian because she does not like meat very much. In her pre-vegetarian days a typical supper included chicken, carrots, baked potato, and white bread. Now she just eats more of everything else to make up for not eating the chicken. What advice would you give her about her diet?

_____

_____

_____

_____

2. Joel loves sweets. He eats candy bars and doughnuts, and he drinks a soft drink every day. He claims that this is a good way to get carbohydrates. Based on what you have learned, what would you tell him?

_____

_____

_____

_____

3. Ming refuses to eat any fats or oils because she is afraid that they will make her gain weight. She does not even butter her bread. Ming thinks a diet without fats or oils is the most healthy diet possible. What would you tell her?

_____

_____

_____

_____

4. After talking about healthy diets with your friends, you start to think more about your own diet. How does the food that you eat provide you with the organic compounds that you need? Is there room for improvement?

_____

_____

_____

_____

Name: _____   Date: _____

Lab 1.2.1A Cells Are Different

QUESTION: How do cells differ?

HYPOTHESIS: _____

_____

EXPERIMENT:

You will need:	• pond water	• 3 toothpicks	• prepared blood cell	• cucumber slice
• microscope	• 4 microscope slides	• 1 leaf	slides	• apple slice
• 2 eyedroppers	• 4 coverslips	• tap water	• yeast	• nonliving things

Steps:
1. Identify all of the parts on your microscope before you begin. Complete **WS 1.2.1A Microscope**.
2. Use a clean eyedropper to place a small drop of pond water on a microscope slide. Holding a coverslip by the edges, carefully lower it over the drop.
3. Move the low-power objective into position over the slide. Use the coarse adjustment knob to move the stage away from the objective as far as it will go. Place the slide on the microscope stage and secure it in place under the stage clips.
4. Looking from the side (not through the ocular lens), lower the body tube using the coarse adjustment knob until the end of the objective lens is just above the coverslip. Do this carefully so you do not crack the slide and possibly damage the objective lens. Look through the ocular lens and slowly turn the coarse adjustment knob until the image is in focus. Once the image is in focus, turn the fine adjustment knob to get a clearer image.
5. Slowly move the slide back and forth. Be careful not to touch the coverslip with your fingers. When you find an organism, stop and observe it. You may need to adjust the focus again to get a clear image.
6. Sketch and describe what you see. Record the magnification (multiply the number on the ocular lens times the number on the objective lens). _____

                 ◯   _____

                              _____

                              _____

                              _____

7. Examine the same organism using greater magnification. Note any details observable now that were not visible under low magnification.

_____

Lab 1.2.1A Cells Are Different

8. Use a toothpick to scrape cells from the leaf of a plant. Place the plant cells on a clean microscope slide. Use a clean eyedropper to place a drop of tap water on the cells. Set a coverslip over the specimen. Follow the same procedure as before to use the microscope. Observe using low magnification first and then high magnification. Record both magnifications. _____

   Sketch and describe the cells under high magnification. _____

9. Prepare slides of other living things and nonliving things. Sketch and label cells from blood, yeast, a nonliving thing, and either the apple or the cucumber.

ANALYZE AND CONCLUDE:

1. How did the organism in the pond water differ from the blood and yeast cells?

_____

2. Describe the differences between the leaf and the apple or cucumber cells. _____

_____

_____

3. Did you observe any cells in the nonliving things? _____

Explain why. _____

4. How were your observations of living things different from those of the nonliving

things? Be specific. _____

_____

_____

5. Compare your observations with your hypothesis. _____

Name: _____ Date: _____

Lab 1.2.2A Plant or Animal?

QUESTION: How do plant cells differ from animal cells? How are they the same?

HYPOTHESIS: _____

_____

EXPERIMENT:

You will need:	• microscope	• metric ruler	• methylene blue or iodine
• onion slice	• eyedropper	• forceps	
• 2 microscope slides	• tap water	• cardboard or cutting board	
• 2 coverslips	• scalpel	• toothpick	

Steps:

1. Observe the onion slice. Describe its physical characteristics. _____

_____

Place the onion slice on the stage. Observe the edge of the slice using the different magnifications available on your microscope. Sketch your observations under each magnification.

Magnification _____   **Magnification** _____   **Magnification** _____

2. Place a small drop of water in the middle of a microscope slide. Separate one onion ring from the slice. Use the scalpel to carefully cut a 1.25 cm piece of onion from the ring. Peel off the thin, transparent membrane from the underside of the onion with the forceps. Place the membrane on the drop of water on the microscope slide. Carefully lower a coverslip over the membrane to secure it into place. Make sure there are no air bubbles. Observe the membrane under different magnifications. Sketch your observations. Label any identifiable parts.

Onion cells

Magnification _____   **Magnification** _____   **Magnification** _____

Lab 1.2.2A Plant or Animal?

With the flat end of a toothpick, gently scrape a saliva sample from the inside of your cheek and smear the sample on a slide. Use the eyedropper to add a very small amount of methylene blue or iodine to make the cells show up better. Be careful with the dye solutions because they can stain your skin and clothes. Place a coverslip over the specimen.

3. Observe the cells under a microscope using the different magnifications available. Sketch your observations and label any identifiable parts.

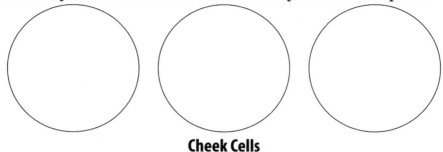

Cheek Cells

ANALYZE AND CONCLUDE:

1. Which specimen was the plant cell? _____

Which was the animal cell? _____

2. How are the cells the same? _____

_____

3. How are the cells different? _____

_____

_____

4. Why are these cells' differences important? _____

_____

_____

_____

5. Compare your hypothesis to the actual results. _____

_____

_____

6. What else could you do to make this experiment more accurate? _____

_____

_____

_____

Name: _____   Date: _____

Lab 1.2.3B Wheat Germ DNA

QUESTION: What does DNA look like?

HYPOTHESIS: _____

_____

EXPERIMENT:

You will need:
- 1 g measuring spoon
- 1 g wheat germ
- 50 mL test tube
- graduated cylinder
- thermometer
- test tube rack
- 20 mL of hot water (50°C–60°C)

- 1 mL of Dawn liquid detergent
- stirring rod
- eyedropper
- paper towels
- 14 mL of 95% ethyl alcohol
- paper clip
- microscope slide
- coverslip

- microscope
- 50% alcohol for storing DNA (optional) (To make 100 mL of 50% alcohol, mix 53 mL of 95% ethyl alcohol with 47 mL of distilled water.)
- sealable container

Steps:
1. Use the 1 g measuring spoon to place 1 g of raw wheat germ in a 50 mL test tube. Place the test tube in the test tube rack.
2. Add 20 mL hot (50°C–60°C) water to the test tube. Stir constantly for 3 minutes.
3. Add 1 mL of Dawn detergent to the test tube. Stir the mixture gently with a stirring rod every minute for 5 minutes. Try not to create foam. If foam appears, use a eyedropper or edge of a paper towel to remove it.
4. Carefully tilt the test tube at an angle and slowly pour 14 mL of ethyl alcohol down the side of the tube to form a layer on top of the water/wheat germ/detergent solution. Do not mix the two layers together.
5. Gently set the test tube in the test tube rack. White, stringy, filmy DNA will begin to appear where the water and alcohol meet. Leave the test tube in the rack for 15 minutes so that the DNA will float to the top of the alcohol.
6. Bend the paper clip to form a hook and carefully collect the DNA. (You can collect more DNA by using the paper clip hook to lift the top of the water layer into the bottom of the alcohol layer.)
7. Place some DNA on a microscope slide and put a coverslip over it. Secure the microscope slide under the stage clips on the microscope.
8. Examine the DNA under low and high magnification. Draw what you observe.

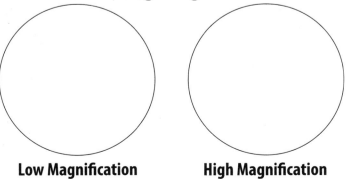

Low Magnification   **High Magnification**

Lab 1.2.3B Wheat Germ DNA

9. If you want to keep the DNA, air dry the DNA strands on a paper towel or store it in 50% alcohol in a sealed container.

ANALYZE AND CONCLUDE:

1. What did you extract from the wheat germ? _____

2. Describe the physical appearance of the extracted DNA. _____

_____

3. Where in the cell is DNA found? _____

4. Think about the process you just used to extract the DNA from the wheat germ.

   a) What was the purpose of the heat (hot water)? _____

_____

   b) What did the detergent do in the extraction process? _____

_____

   c) What was the purpose of using the alcohol? _____

_____

5. What are some possible errors in this experiment? _____

_____

_____

6. What could you do to improve this experiment? _____

_____

_____

7. Compare your hypothesis to the results. _____

_____

8. Comment on the amount of information that is packed inside one strand of DNA. What does this tell you about God's power? _____

_____

_____

9. What have you learned from this experiment? _____

_____

_____

Name: _____ Date: _____

Lab 1.2.4A The Enzyme Effect

QUESTION: How do enzymes affect reactions?

HYPOTHESIS: _____

_____

EXPERIMENT:

You will need:	• bowl	• cold water	• 4 kiwi chunks
• marking pencil	• measuring cup	• 4 fresh pineapple chunks	• 4 orange slices (cut in half)
• 6 beakers	• boiling water	• 4 canned pineapple chunks	• refrigerator
• 1 package of gelatin	• spoon	• 4 apple chunks	

Steps:
1. Thoroughly clean and dry all the beakers. With the marking pencil, label the beakers *Gelatin*, *Fresh Pineapple*, *Canned Pineapple*, *Apple*, *Kiwi*, and *Orange*.
2. Prepare the gelatin in the bowl according to the directions on the package. (Gelatin is a protein that solidifies as it absorbs water.)
3. Fill each beaker half full of the liquid gelatin. Use the measuring cup to remove the gelatin from the bowl and pour it into the beakers.
4. Use your fingers to add four chunks of each fruit to the corresponding marked beakers. Leave the control beaker without any fruit. Wipe your hands off before handling each fruit. Be careful not to allow the juice of any fruit into another beaker or your results may be inaccurate. Do not stir the contents.
5. Refrigerate the beakers overnight.
6. The next day, check to see which contents solidified. Record your observations.

Beaker Contents	**Observations**
Gelatin	
Fresh Pineapple and Gelatin	
Canned Pineapple and Gelatin	
Apple and Gelatin	
Kiwi and Gelatin	
Orange and Gelatin	

Lab 1.2.4A The Enzyme Effect

ANALYZE AND CONCLUDE:

1. What was the independent variable in this experiment? _____

2. What was the dependent variable? _____

3. In which conditions did the gelatin set? _____

   _____

4. In which conditions did the gelatin remain a liquid? _____

   _____

5. Why did those particular fruits keep the gelatin from solidifying? _____

   _____

6. All living cells produce enzymes of some type. Since gelatin is a protein, which fruit contains enzymes that break down proteins? _____

7. Which fruit does not contain the enzymes that break down proteins? _____

   _____

8. Compare your hypothesis to the results. _____

   _____

   _____

Research to find the answers to the following questions:

9. What type of enzyme breaks down protein? _____

10. Why did the canned pineapple allow the gelatin to solidify when the fresh pineapple did not? _____

11. What can you do to the fresh pineapple that would allow the gelatin to solidify?

   _____

12. If you wanted to make a marinade that would help tenderize meat, would you use fresh pineapple or canned pineapple? Why? _____

   _____

13. What other fruits would not be good to put in gelatin? _____

14. What other fruits would work well with making gelatin? _____

15. Does freezing the fruit inactivate the protease enzymes? _____

Name: _____ Date: _____

Lab 1.2.5A Cell Size

QUESTION: Why must cells be so small?

HYPOTHESIS: _____

_____

EXPERIMENT:

You will need:	• triple beam balance	• water
• 100 sugar cubes	• white glue	• graph paper
• metric ruler	• 4 large beakers	• stopwatch

Steps:
1. Determine the mass of a sugar cube and fill in the chart for the 1-cube cell.
2. Use the metric ruler to measure the sides of a sugar cube. Fill in the chart for the area and total surface area.
3. Build a 2 × 2 × 2 cube cell by gluing 8 sugar cubes together. Determine the mass (in grams), area (in centimeters), and total surface area and fill in the chart for the 8-cube cell.
4. Build a 3 × 3 × 3 cube cell by gluing 27 sugar cubes together. Determine the mass, area, and total surface area and fill in the chart for the 27-cube cell.
5. Build a 4 × 4 × 4 cube cell by gluing 64 sugar cubes together. Determine the mass, area, and total surface area and fill in the chart for the 64-cube cell.

Sugar-Cube Cell Data Chart

	Mass	Area of One Face ($l \times h$)	Total Surface Area ($l \times h \times 6$)	Volume ($l \times w \times h$)
1-Cube Cell				
8-Cube Cell				
27-Cube Cell				
64-Cube Cell				

6. On the graph paper, use your data to create a graph comparing the ratio of total surface area to volume and the ratio of total surface area to mass for each cube.
7. Predict how long it will take for each sugar-cube cell to dissolve in a container of water. Record your predictions on the chart.

Dissolving Times

	Prediction	Actual Time
1-Cube Cell		
8-Cube Cell		
27-Cube Cell		
64-Cube Cell		

Lab 1.2.5A Cell Size

8. Submerge each cube in a beaker of water. Make sure the cubes are completely covered by the water. Measure the time it takes for each cube to dissolve. Fill in the chart.

ANALYZE AND CONCLUDE:

1. What happens to the ratio of total surface area to volume as the cell grows in size?

_____

_____

2. Will the cell membrane of a large cell or a small cell be more effective at supplying food throughout the cell's cytoplasm? Why? _____

_____

3. What happens to the ratio of the total surface area to mass ratio as the cell grows in size? _____

_____

4. Which can better feed the cell's cytoplasm—the cell membrane of a cell with high mass or the cell membrane of a cell with low mass? Why? _____

_____

_____

5. Was your hypothesis correct? Explain. _____

_____

_____

6. Why do cells need to be so small? _____

_____

_____

Name: _____ Date: _____

Lab 1.2.5B Osmosis

QUESTION: How does osmosis work?

HYPOTHESIS: _____

_____

EXPERIMENT:

You will need:	• distilled water	• triple beam balance
• 6 beakers	• 2 baby carrots (same size and mass)	• forceps
• marking pencil	• 2 grapes (same size and mass)	• paper towels
• salt	• 2 raisins (same size and mass)	

Steps:
1. Label 3 beakers *Saltwater* and 3 beakers *Distilled Water*.
2. Fill the beakers half full of distilled water. Dissolve as much salt as possible in the 3 beakers labeled *Saltwater*.
3. Record your observations of the physical characteristics of the carrots, grapes, and raisins in the *Before* column.

Observations

	Carrot in Saltwater	Carrot in Distilled Water	Grape in Saltwater	Grape in Distilled Water	Raisin in Saltwater	Raisin in Distilled Water
Before						
After						

4. Use the triple beam balance to measure and record the mass of one carrot. Use the forceps to place it in the saltwater. Measure and record the mass of the second carrot on the Data Chart and place it in the distilled water. Clean the balance after each use. Repeat this procedure for the grapes and the raisins.
5. Predict what you think will happen to the shape, mass, and flexibility of each carrot, grape, and raisin. Record your predictions.
6. After 24 hours, remove the carrot from the saltwater. Do not stab or poke the carrot with the forceps while removing it. Pat the carrot dry with a paper towel. Record your observations on the chart. Measure and record the carrot's mass on the Data Chart.
7. Repeat Step 7 for the second carrot and the grapes and raisins.
8. Record the difference between the masses on the Data Chart.

Lab 1.2.5B Osmosis

Predictions

Carrot in Saltwater	Carrot in Distilled Water	Grape in Saltwater	Grape in Distilled Water	Raisin in Saltwater	Raisin in Distilled Water

Data Chart

	Carrot in Saltwater	Carrot in Distilled Water	Grape in Saltwater	Grape in Distilled Water	Raisin in Saltwater	Raisin in Distilled Water
Day 1 Mass						
Day 2 Mass						
Difference						

ANALYZE AND CONCLUDE:

1. What happened to the carrot, grape, and raisin in saltwater? _____

_____

2. What happened to the carrot, grape, and raisin in distilled water? _____

_____

3. What do you think caused the changes in mass in relation to osmosis? _____

_____

_____

_____

4. Compare the results to your hypothesis. _____

_____

5. What do you predict will happen if you put the items from the saltwater into the

distilled water and the items from the distilled water into the saltwater?

_____

_____

6. Why should you not drink saltwater no matter how thirsty you are? _____

_____

_____

Microscope

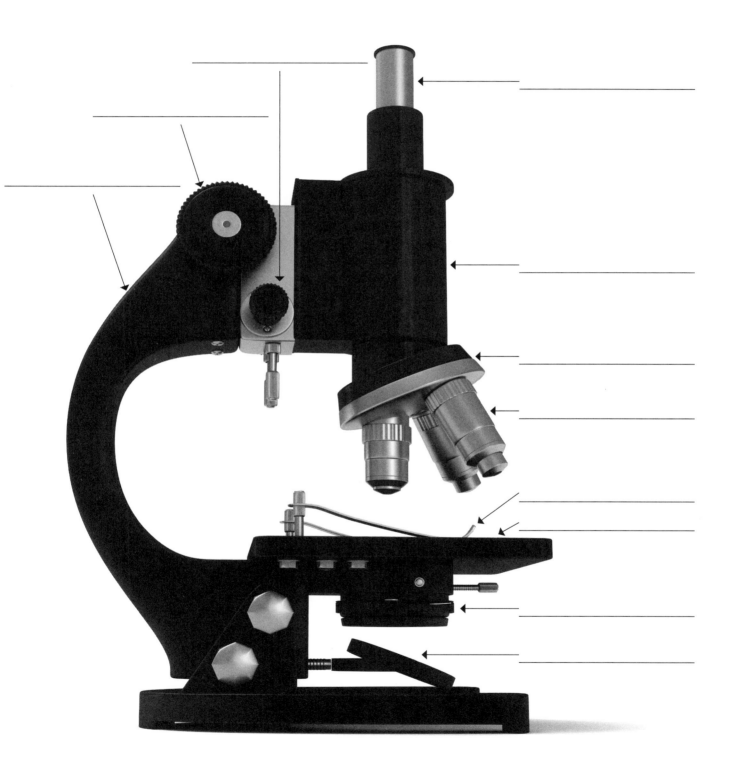

Animal Cell   WS 1.2.2A

1. Label the parts of the cell.

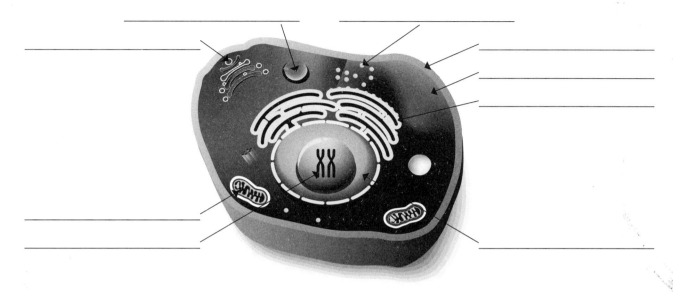

Describe the function of the following organelles.

2. nucleus _____

3. mitochondrion _____

4. Golgi apparatus _____

5. ribosome _____

6. endoplasmic reticulum _____

7. chromosome _____

8. lysosome _____

9. cytoplasm _____

10. nuclear membrane _____

Plant Cell

1. Label the parts of the cell.

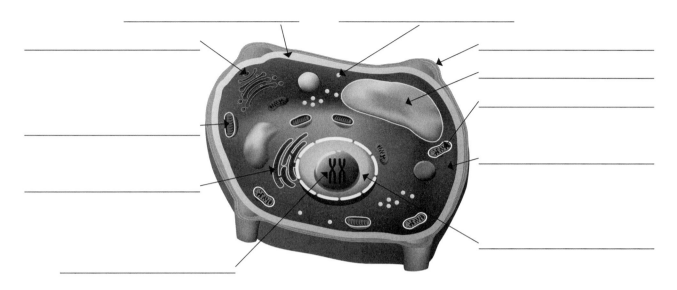

Describe the function of the following organelles.

2. chloroplasts _____

3. cell wall _____

4. vacuole _____

5. cell membrane _____

Research and write your answer. Include your source.

6. Do plant cells have lysosomes? _____

_____

_____

_____

Cell Crossword

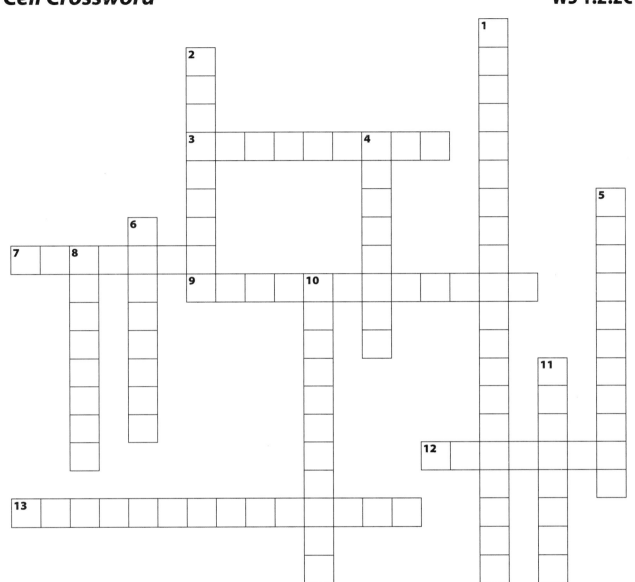

Cell Crossword continued   **WS 1.2.2C**

Across

3. Any structure inside a cell that performs a particular function is called an _____.

7. The _____ is the cell's control center.

9. The _____ produce energy for the cell.

12. A _____ is a storage structure inside a plant cell.

13. This organelle modifies and transports lipids and proteins.

Down

1. The cell's transportation system is called the _____ _____.

2. This is the watery fluid in the cell.

4. This organelle releases chemicals that aid in the cell's digestive process.

5. The _____ direct the cell's growth and reproduction.

6. The cell _____ controls what goes in and out of the cell.

8. A plant cell has this thick layer, but an animal cell doesn't.

10. Chlorophyll is found within a _____.

11. The _____ provides the cell with proteins.

Cell Tour

You are part of a research team made up of an oceanographer and a microbiologist. The government has called you into their research laboratories to send you on an important medical mission. You and your partner will be placed into a small submarine, shrunk down to microscopic size, and then injected into a living cell that is about to become diseased with cancer. Your mission is to find the strand of DNA that is about to cause the cancer and bring a piece of it back for researchers to study. As you travel through the cell, identify your location and write it in the space provided. If you make three mistakes in identifying the various structures as you travel through the cell, your power source will become inactive and you will float aimlessly through swirling cytoplasm for the rest of your days. Be careful—and good luck.

The Adventure

You and your submarine have been shrunk to microscopic size and then injected into the living cell. Your sub is floating in a vast sea of a thick, jellylike substance that fills the entire cell. Thousands of brightly colored crystals drift in the currents around your submarine, like sparkling jewels in a thick soup. In order to make the appropriate navigational and steering adjustments, you need to find out what this substance is. What are you floating in?

_____

You receive a report from the research lab that the location of the cell structure containing the DNA is near another structure that is releasing large amounts of energy. You will use your heat sensors to find this source of energy. What structure are you looking for?

_____

After some fancy piloting, you see the cell structure that is giving off the energy looming in front of you. It is a giant oval. The thick yellow substance surrounding it has been turned bright orange from the heat created by its energy. To find the DNA, you must now look for a giant sphere located in the center of the cell. This sphere acts like the cell's brain. What structure are you looking for?

_____

You find the giant sphere but discover that it is enclosed by a membrane stretching across its entire surface. The membrane regulates what goes in and out of the sphere—including you. Identify this membrane so that you can enter the large sphere.

_____

You enter the sphere and are traveling toward its center when you realize that your submarine is entangled by long, black, threadlike structures massed together into a large ball. As they wrap themselves around your submarine, they slowly begin to crush the hull of your ship. Inside you hear the metal of your ship begin to creak and grind as it tries to withstand the pressure of the crushing threads. If you know what the threadlike structures are, you can inject them with the right chemicals that will cause them to release you. What are they?

_____

The black threads release you when you inject the proper chemical into them. Finally you locate the portion of abnormal DNA that will cause the cell to become cancerous. Special cutting tools emerge from your submarine, snip off the section of DNA, and store it away in the hull. You are on your way out of the structure when a strong current in the fluid pulls you down into a long, winding tunnel. This tunnel is part of an interconnecting transport system inside of the cell. It is pitch-black inside. You turn your lights on to see the familiar brightly colored crystals of sugar, salt, and other materials flowing in the tunnel's current with you. Your lights catch an eerie glow on the dark red walls of the tunnel, which hurl past you with frightening speed. Identify the cell structure you are traveling through.

_____

As you pilot yourself out of the network of tunnels, you become trapped inside a small sphere that is attached to the tunnel. The sphere is producing bright red chains of protein. Identify what structure has you trapped so that you know what cutting tool to use to slice your way out of it.

_____

Again you are trapped by an organelle. This one is a large sac. Once you are inside, powerful acids immediately begin corroding through the hull of your submarine. In order to neutralize these powerful enzymes and escape, you must identify the organelle that is trapping you.

_____

You cut your way out of the spherical structure, and once again you are floating in the thick, sea of jelly inside the cell. Now you must find your way out of the cell so that the lab researchers can identify your submarine and restore it and you to the proper size. You are looking for the barrier between the cell and the outside world. Identify the structure you are looking for.

_____

DNA

Sugar

Sugar

Phosphate

Phosphate

Adenine

Sugar

Sugar

Phosphate

Phosphate

Adenine

Sugar

Sugar

Phosphate

Phosphate

Adenine

Sugar

Sugar

Phosphate

Phosphate

Thymine

Sugar

Sugar

Phosphate

Phosphate

Thymine

Sugar

Sugar

Phosphate

Phosphate

Thymine

Cytosine

Cytosine

Cytosine

Guanine

Guanine

Guanine

Mitosis

Draw the stages of a cell dividing through interphase, mitosis, and cytokinesis.

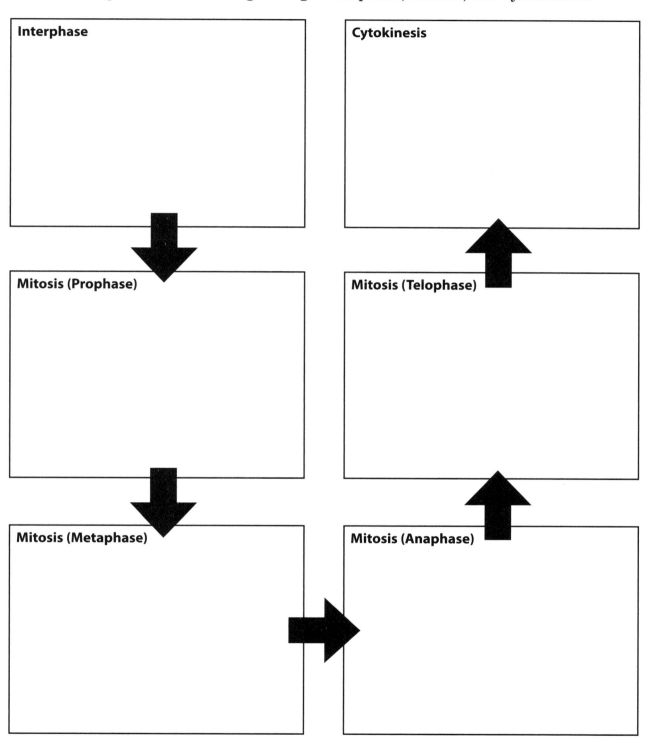

Interphase

Cytokinesis

Mitosis (Prophase)

Mitosis (Telophase)

Mitosis (Metaphase)

Mitosis (Anaphase)

Lab 1.3.2A Taxonomy

QUESTION: How does classification work?

HYPOTHESIS: _____

_____

EXPERIMENT:

You will need:	• 15–20 buttons	• 24–30 pieces of yarn	• 30 index cards

Steps:

1. Label an index card with the word *Buttons*.
2. Decide how to divide the group of buttons into two or three smaller groups. Write the group name on a card. For example, if you were dividing up different types of candy, the groups might be Chocolate and Non-Chocolate. Use yarn to connect the *Button* card to each new group so that they are clearly branching off from the main group.
3. Examine each smaller group and divide the buttons into new categories based on their characteristics. For example, candy in the non-chocolate group could be further divided as Hard and Soft. Write the new category names on index cards. Use yarn to connect the new categories to the secondary groups.
4. Analyze the buttons in each group. What characteristics do you see? How can they be grouped or divided again? Continue dividing the buttons into subcategories. Label each subcategory using an index card. Use yarn to connect each subcategory of buttons it came from.
5. When the buttons have all been categorized, write the various characteristic names on index cards. For example, the non-chocolate candy could be have subcategories based on flavor, texture, shape, or color, such as Chewy, Flat, Round, Sour, or Sweet. Use yarn to connect the cards so that the branches of each category of buttons is clear.

ANALYZE AND CONCLUDE:

1. Was it difficult to classify the buttons?

2. Did some of your classmates classify the buttons differently than you did? Why?

3. If you could reclassify some of the buttons, how would you do it?

4. Considering the vast number of species in the world, why do you think it is important for scientists to classify organisms?

Classify

Common name of organism _____

1. Use the Internet or a field guide to research the scientific classification of your organism. List at least three other organisms in each scientific category.

Domain _____

other organisms _____

_____

Kingdom _____

other organisms _____

_____

Phylum _____

other organisms _____

_____

Class _____

other organisms _____

_____

Order _____

other organisms _____

_____

Family _____

other organisms _____

_____

Genus _____

other organisms _____

_____

Species _____

2. What do you find interesting about your organism? _____

_____

_____

Match Me

Matching

_____ **1.** house cat

_____ **2.** cactus

_____ **3.** yeast

_____ **4.** corn

_____ **5.** penicillin fungus

_____ **6.** Scotch pine

_____ **7.** white pelican

_____ **8.** dog

_____ **9.** lion

_____ **10.** horse

_____ **11.** human

_____ **12.** box turtle

a. *Panthera leo*

b. *Terrapene carolina*

c. *Pelecanus erythrorhynchos*

d. *Canis familiaris*

e. *Equus caballus*

f. *Homo sapiens*

g. *Felis catus*

h. *Kluyveromyces lactis*

i. *Zea mays*

j. *Cereus giganteus*

k. *Pinus sylvestris*

l. *Penicillium chrysogenum*

Lab 2.1.3A Virus Infection

QUESTION: How do viruses spread?

HYPOTHESIS: _____

_____

EXPERIMENT:

You will need:	• prepared cups

Steps:
1. Choose a cup of fluid.
2. Examine the fluid in your cup and write down a description of what the fluid

   looks like. _____

   _____

3. Share the fluid in your cup with another student by pouring all of it into his or
   her cup. Then have that student pour half of the fluid back into your cup. Write
   down the number of that student's cup below. Repeat this step with two other
   classmates.

   Cup Numbers   _____   _____   _____

4. After the teacher puts phenolphthalein solution in your cup, describe the fluid in

   your cup. _____

   _____

ANALYZE AND CONCLUDE:

1. Did the fluid in your cup change appearance? _____

_____

2. Could you tell if a cup was infected before the indicator was added? _____

3. What percentage of students ended up being infected? _____

_____

4. Why is it important to know how viruses spread? _____

_____

5. What factors could influence how a virus spreads? _____

_____

6. How can studying viruses help prevent the spread of disease? _____

_____

Virus Research

Choose a virus that affects humans, animals, or plants. Record the name of the virus below.

Virus _____

Draw the virus, label its parts, and include size information.

Write at least five facts about the virus, including the type of organism it affects and what disease it causes.

1. _____
_____
_____

2. _____
_____
_____

3. _____
_____
_____

4. _____
_____
_____

5. _____
_____
_____

Name: _____ Date: _____

Virus Attack!

Draw the process of a bacteriophage attack and reproduction within a bacterium. Write captions to explain what is happening.

Viral Vocabulary WS 2.1.3A

Across

4. The _____ virus causes AIDS.

10. Once a virus or its nucleic acid has invaded a host cell, _____ has taken place.

11. If an animal bites a person, that animal should be tested for _____.

Down

1. Even a bacterium can be infected by a virus known as a _____.

2. Antibiotics will not cure a cold because it is caused by a _____.

3. Viruses can reproduce because they have _____ _____ such as RNA or DNA.

5. Acquired _____ Syndrome destroys the body's immune system.

6. Viruses can make protein because they contain _____ _____.

7. This is the type of microscope you would use to view viruses.

8. If you had _____, you would have achy muscles, fever, and the chills.

9. HIV is a _____ because its genetic material is in the form of RNA.

Name: _____ Date: _____

Lab 2.2.1A Bacteria Structure

QUESTION: How does the structure of bacteria vary?

HYPOTHESIS: _____

_____

EXPERIMENT:

You will need:	• plain yogurt	• eyedropper	• 2 coverslips	• cyanobacteria
• 2 toothpicks	• water	• 2 microscope slides	• microscope	culture

Steps:
1. Use a toothpick to place a tiny spot of plain yogurt on a microscope slide.
2. Add a very small drop of water and mix it in with the yogurt.
3. Place a coverslip over the mixture.
4. Observe the yogurt through the microscope under low power and then under high power. Focus on the bacterial cells. You may have to reduce the amount of light in order to see the small cells. These bacterial cells are lactobacilli.
5. Sketch several cells under each power. Label the magnification below each drawing.

Magnification _____   Magnification _____

6. Describe the structure and shape of the lactobacilli. _____

_____

_____

_____

_____

7. Prepare a microscope slide with cyanobacteria.

Lab 2.2.1A Bacteria Structure

8. Examine the cells under low and then high power. Adjust the light, if necessary. Sketch your observations of the cyanobacteria. Label the magnification below each drawing.

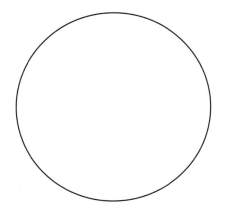

Magnification _____

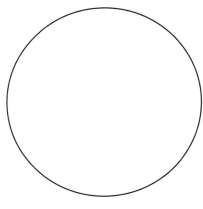

Magnification _____

9. Describe the structure and shape of the cyanobacteria. _____

_____

ANALYZE AND CONCLUDE:

1. Name one similarity between the lactobacilli and the cyanobacteria. _____

_____

2. Name two differences between the lactobacilli and the cyanobacteria. _____

_____

_____

_____

3. Using your observations, how might the cyanobacteria obtain energy? _____

_____

4. Think about what you know about lactobacilli and infer whether it is a helpful or

harmful type of bacteria. Explain your answer. _____

_____

_____

_____

5. Why is cyanobacteria not classified as a protist or a plant? _____

_____

Lab 2.2.2A Bacteria Growth

QUESTION: Will the same kind of bacteria grow in different conditions?

HYPOTHESIS: _____

_____

EXPERIMENT:

You will need:	• 3 wide-mouthed	• 5 tsp sugar	• water	• 2 eyedroppers
• 5 tsp powdered, unflavored gelatin	quart jars with lids	• measuring spoon	• 2 microscope slides	• microscope
• 2 cups garden soil	• stirring rod	• 9 tsp mineral fertilizer	• 2 microscope coverslips	
	• marking pencil			

Steps:

1. Pour 5 tsp of powdered gelatin and 1 cup of garden soil into a wide-mouthed jar. Mix with a stirring rod. With the marking pencil, label the jar *Protein*.

2. In a wide-mouthed jar, use a stirring rod to mix 5 tsp of sugar with 1 cup of garden soil. Label the jar *Carbohydrates*.

3. Add 9 tsp of mineral fertilizer to a wide-mouthed jar half filled with water. Screw the lid on tightly, and shake the mixture.

4. Divide the liquid fertilizer mixture between the Protein jar and the Carbohydrate jar. Put the lids on. Gently shake the jars.

5. Review your hypothesis. Explain why you answered the way you did. _____

_____

_____

6. Let the jars stand for two weeks. Observe the jars daily and record your observations on a chart.

Week 1

Days	Protein Observations	Carbohydrate Observations
1		
2		
3		
4		
5		
6		
7		

Lab 2.2.2A Bacteria Growth

Week 2

Days	Protein Observations	Carbohydrate Observations
1		
2		
3		
4		
5		
6		
7		

ANALYZE AND CONCLUDE:

1. Prepare a microscope slide with a small bacteria sample from the Protein jar. Observe it under low and high magnification. Draw your observations. Label the magnification for each drawing.

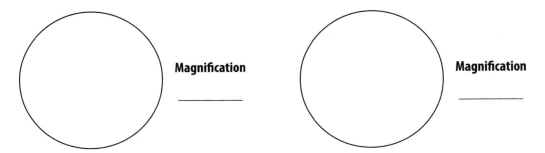

Magnification

_____

Magnification

_____

2. Prepare a microscope slide with a small bacteria sample from the Carbohydrate jar. Observe it under low and high magnification. Draw your observations. Label the magnification for each drawing.

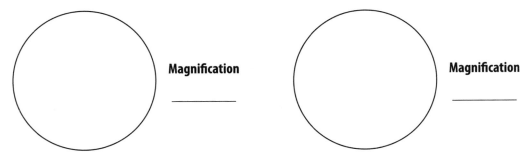

Magnification

_____

Magnification

_____

3. Did the same kind of bacteria grow in each mixture? How can you tell? _____

_____

4. How might this knowledge relate to preserving foods? _____

Name: _____ Date: _____

Lab 2.2.4A Bacteria and Yogurt

QUESTION: How do bacteria help make yogurt?

HYPOTHESIS: _____

_____

EXPERIMENT:

You will need:	• 10 mL sugar water	• 10 mL water	• 50 mL of milk
• 5 paper cups	• 10 mL lemon juice	• 10 toothpicks	• 5 spoons
• marking pencil	• 10 mL coffee	• 2 pH paper strips	• yogurt container
• graduated cylinder	• 10 mL vinegar	• pH color scale	ingredient list

Steps:
1. Use the marking pencil to label the paper cups *Sugar*, *Lemon*, *Coffee*, *Vinegar*, and *Water*. Pour 10 mL of the appropriate liquid into the cups.
2. Dip a toothpick into each cup (using a clean toothpick each time), and put a dot of the liquid on a pH strip. Make sure to keep the dots far enough apart so they do not run together. Compare the color of each to the scale on the pH paper box. Record the pH levels in the pH *Before* section of the chart below.
3. Add 10 mL of milk to each cup. Stir each mixture with a clean spoon. Do not cross contaminate the cups.
4. Dip a clean toothpick into each cup and put a drop of each liquid on a pH strip. Record the pH level in the pH *After* section of the chart.

	Sugar	**Lemon**	**Coffee**	**Vinegar**	**Water**
pH–Before					
pH–After					

ANALYZE AND CONCLUDE:

1. Which substance has the lowest pH level? _____

_____

2. Which substances had an acid pH? _____

_____

3. Describe what happens to the milk when it is added to vinegar. _____

_____

4. Describe how the milk reacts with other substances. _____

_____

_____

_____

Lab 2.2.4A Bacteria and Yogurt

5. How does a liquid's pH relate to its reaction with milk? _____
_____
_____

6. Examine the ingredient list on the yogurt container. What is the main
ingredient? _____
_____

7. What do you think "live active cultures" are? _____
_____

8. Explain why yogurt, a milk product, is so thick. _____
_____

9. What do you think is the role of the live active cultures in the yogurt? _____
_____

10. Compare your hypothesis to the results. _____
_____

11. Research to find out at least three types of bacteria used to make yogurt. List the
genus and species of each. _____
_____

Lab 2.2.5A Harmful Bacteria

QUESTION: Where am I most likely to encounter harmful bacteria?

HYPOTHESIS: _____

_____

EXPERIMENT:

You will need:	• 10 sterilized petri dishes	• various foods and objects	• marking pencil
• 7 sterilized cotton swabs	containing a nutrient medium	to test for bacteria	

Steps:

1. Do not open the petri dishes or expose the cotton swabs to air until you are ready to use them. Always wash your hands after handling the dishes. Do not open the dishes when your experiment is complete. Give them to your teacher to dispose of. Use the marking pencil to label one petri dish with the word *Control*. Do not open this dish.
2. Expose one petri dish to the air for 5 minutes. Seal it. Then label it *Air—5 min*.
3. Expose a different petri dish to the air for 30 minutes. Seal it. Then label it *Air—30 min*.
4. Choose seven different objects to test for bacteria. (Suggestions: desktop, doorknob, various foods, spoon with which someone has eaten, pencil, inside surface of cheek, mouthpiece of musical instrument, insect, washed hand, unwashed hand, faucet handle. You may also choose to cough or breathe on the nutrient medium of a petri dish.)
5. Test one of your chosen objects by rubbing a sterilized cotton swab across its surface. Quickly open the petri dish cover and rub the swab across the solid nutrient medium. Quickly remove the swab and close the dish immediately. Seal and label the dish with the name of the chosen item.
6. Repeat Step 5 with the other six objects. Label the petri dishes accordingly.
7. Predict which surface sample will yield the most bacteria. Record your

   prediction. _____
   Predict which surface sample will yield the least bacteria. Record your

   prediction. _____
8. Turn the dishes upside down; store them in a dark, warm place for four days.
9. Observe the dishes each day and record your observations on the data chart.

Note: This procedure also generates mold. The mold colonies will be fuzzy. The bacteria will form small, often colored, soft, pasty-looking colonies.

Lab 2.2.5A Harmful Bacteria

Observations	Day 1	Day 2	Day 3	Day 4
Control				
Air–5 min.				
Air–30 min.				

ANALYZE AND CONCLUDE:

1. Where did the most bacteria grow? _____

2. Where did the least bacteria grow? _____

3. Did any bacteria grow on the control dish? _____

4. Compare the results to your hypothesis. _____

_____

5. Were you surprised by the amount of surfaces that produced bacteria? Explain.

_____

_____

6. What are some practical applications that you could draw from this experiment?

_____

_____

Bacteria Business

Circle the correct word to complete the sentence.

1. Bacteria are unicellular organisms without a _____.

nucleus   cell membrane

2. Some bacteria have a thin, whiplike structure called a _____ that helps them move.

bud   flagellum

3. The offspring produced by asexual reproduction is _____ to the parent.

identical   not identical

4. Some bacteria reproduce by splitting into two identical daughter cells. This is called _____.

binary fission   budding

5. Some bacteria reproduce through a method in which a small part of the parent bacterium develops into an independent bacterium. This is called _____.

binary fission   budding

6. Some bacteria survive hot, dry conditions by forming _____.

endospores   sporozoans

7. Some bacteria can make their own food.

autotrophs   heterotrophs

8. Some bacteria must obtain their food from an outside source.

autotrophs   heterotrophs

9. Many bacteria get their energy by breaking down the remains of dead organisms or animal wastes. These are _____.

decomposers   endospores

10. Bacteria that are round in shape are called _____.

cocci   bacilli

11. A relationship in which one organism lives on, near, or inside another organism and at least one of the organisms benefits.

symbiosis   cooperation

12. Drugs used to kill harmful bacteria are _____.

vaccines   antibiotics

Bacteria and Archaea   **WS 2.2.6A**

Fill in the diagram with details about bacteria and archaea.

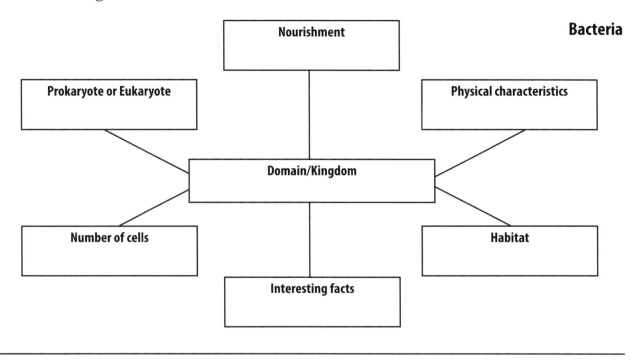

Bacteria

- Nourishment
- Prokaryote or Eukaryote
- Physical characteristics
- Domain/Kingdom
- Number of cells
- Habitat
- Interesting facts

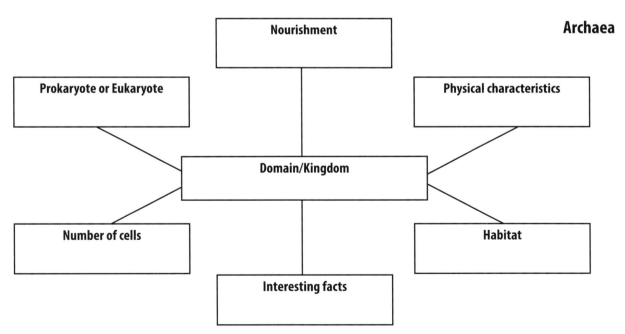

Archaea

- Nourishment
- Prokaryote or Eukaryote
- Physical characteristics
- Domain/Kingdom
- Number of cells
- Habitat
- Interesting facts

Write a sentence comparing bacteria and archaea. Include two major features. _____

_____

_____

Name: _____   Date: _____

Lab 2.3.1A Mushroom Reproduction

QUESTION: How are mushrooms designed to reproduce?

HYPOTHESIS: _____

_____

EXPERIMENT:

You will need:	• 3 mushrooms, each of a	• 2 large bowls	• metric ruler
• 1 knife	different species	• hair spray	
• 3 piece of black paper	• 3 pieces of white paper	• hand lens	

Steps:

1. Using the knife, cut the stems from the mushrooms. Cut each mushroom top in half. (Note: If you are using wild mushrooms for this lab, wash your hands well following this activity.) With the cut side facing toward you, draw the three mushrooms. Write the type of the mushroom under each box.

Mushroom 1	**Mushroom 2**	**Mushroom 3**
_____	_____	_____

2. Place half of each mushroom gill side down on the piece of black paper; place the other halves on a piece of white paper.
3. Move the mushroom halves close enough together so that they can be covered with bowls, but keep the mushrooms at least 2.5 cm away from each other.
4. After 24 hours, remove the bowls and mushrooms. Describe what you see. _____

_____

_____

5. Using hair spray, spray the spore prints made on both pieces of paper to make a permanent print. Hold the can at least 45 cm from the print as you spray to avoid scattering the spores.
6. Examine each print with a hand lens.

Lab 2.3.1A Mushroom Reproduction

ANALYZE AND CONCLUDE:

1. Comment on the number of spores produced by each mushroom. How many

mushrooms can be produced by the spores of one mushroom? _____

_____

_____

2. Why was it necessary to use two different colors of paper? _____

_____

_____

3. How are the spore prints similar? _____

_____

_____

How are they different? _____

_____

_____

4. If you had lifted the mushrooms from the paper a few minutes after placing them
there, you would not have seen many spores. Why is it a good thing that the spores

fall slowly? _____

_____

_____

5. Research the three types of mushrooms you observed. List the scientific name
(genus and species) for each one. Write one fact about each one that makes it
different from the others.

a. _____

_____

b. _____

_____

c. _____

_____

Name: _____ Date: _____

Lab 2.3.2A Yeast

QUESTION: How does yeast grow?

HYPOTHESIS: _____

_____

EXPERIMENT:

You will need:	• 500 mL warm water	• 2 eyedroppers	• microscope
• 4 tsp yeast	• graduated cylinder	• 3 microscope slides	
• 2 small bottles with lids	• 1 tsp molasses	• 3 coverslips	

Steps:
1. Place 2 tsp of yeast into a small bottle with 250 mL of warm water. Put on the lid and shake the bottle. In the second small bottle, mix yeast, warm water, and 1 tsp of molasses. Put on the lid and shake the bottle.
2. Place a drop of yeast water from the first bottle on a slide and observe it under the microscope. Draw what you see.
3. Estimate the number of yeast cells in the drop. _____
4. Concentrate on one yeast cell. Record your observations. _____

   _____

   _____
5. Let mixtures stand for 30–60 minutes.
6. Place a drop of the yeast/molasses mixture from the second bottle on a slide and observe it under a microscope. Estimate how many yeast cells are in that drop.

   Record your estimate. _____
7. Repeat Step 6 for the yeast/water mixture from the first bottle. Estimate how

   many yeast cells are in that drop now. Record your estimate. _____

ANALYZE AND CONCLUDE:

1. How does yeast reproduce? _____

_____

2. What are the best conditions for yeast to reproduce? _____

_____

Lab 2.3.2B Mold

QUESTION: What does mold look like close up?

HYPOTHESIS: _____

_____

EXPERIMENT:

You will need:	• moldy bread	• Roquefort or blue	• microscope slide	• microscope
• hand lens	• thin needle	cheese	• coverslip	

Steps:

1. Use a hand lens to examine the mold on the bread. Sketch and describe what you observe.

2. Use a thin needle to take mold from a vein of Roquefort or blue cheese.

3. Place the sample on a slide, and examine it under a microscope. Sketch and describe what you see.

Lab 2.3.2B Mold

ANALYZE AND CONCLUDE:

1. What are the "knobs" on the bread mold, and what do they do? _____

_____

2. Do the same kinds of mold grow on different kinds of food? How can you tell?

_____

_____

3. On a separate piece of paper, pose a question about mold, and design an experiment to test your hypothesis.

Lab 2.3.3A How Does Mold Grow?

QUESTION: What conditions are best for mold to grow?

HYPOTHESIS: _____

_____

EXPERIMENT:

You will need:	• spray bottle	• marking pencil
• 3 pieces of one kind of bread	• water	
• 6 wide-mouthed jars with lids	• masking tape	

Steps:
1. Tear up the 3 pieces of bread, and place equal amounts in each jar.
2. Moisten the bread by spraying it with water in 3 of the jars.
3. Label these 3 jars *Wet/Light, Wet/Dark*, and *Wet/Cool*.
4. Label the other 3 jars *Dry/Light, Dry/Dark*, and *Dry/Cool*.
5. Leave all 6 jars uncovered overnight.
6. Tightly screw on the lids of the jars. Place the "light" jars on a window sill, place the "dark" jars in a cabinet or closet, and place the "cool" jars in a refrigerator.
7. Which jar do you think will grow the most mold? Record your prediction.

_____

8. Which jar do you think will grow the least mold? Record your prediction.

_____

9. Observe the jars daily for seven days, and record your observations.

Mold Growth

	Wet/Light	Wet/Dark	Wet/Cool	Dry/Light	Dry/Dark	Dry/Cool
Day 1						
Day 2						
Day 3						
Day 4						
Day 5						
Day 6						
Day 7						

Lab 2.3.3A How Does Mold Grow?

ANALYZE AND CONCLUDE:

1. Which jar grew the most mold? _____

Which grew the least? _____

2. Compare your hypothesis to the actual results. _____

_____

3. What do you predict would happen if you placed a sample in a very hot or a very

cold room? _____

_____

4. What do you predict would happen if you completely immersed the bread in water?

_____

_____

5. What do you think would happen if you repeated this experiment using toast?

Why? _____

_____

6. What differences would you expect to see if you were to use bread with
preservatives in some jars and bread without preservatives in other jars?

_____

_____

7. On a separate piece of paper, pose a question and design an experiment to test the
hypothesis you wrote in Exercise 6. Include step-by-step instructions and some
"analyze and conclude" questions.

8. How can you apply what you have learned about mold growth? _____

_____

Name: _____ Date: _____

Fungal Facts

1. Write down three numbers between 1 and 10. _____ _____ _____
2. Circle the correct answers below.
3. Then, perform all eight of the following mathematical operations for each number you chose above. The chosen number is the factor in the first equation. The answer for each problem becomes the next factor in the following equation. If you get the same answer each time you complete all operations, you have answered each question correctly.

- These threadlike filaments of fungi produce enzymes that break down dead or living organisms.

  hyphae ($\times$ 2) flagellum ($\times$ 3)

- This mass of hyphae makes up a fungi's growing structure.

  mycelium (+ 9) gills (+ 7)

- Fungi use these reproductive cells.

  sperm (+ original number) spores (- 6)

- Molds have this property in common.

  fuzzy (+ 11) photosynthesis (+ 3)

- All yeast

  reproduce by fission. ($\times$ 4) are unicellular. ($\times$ 6)

- All mushrooms have

  gills. ($\div$ 4) a stalk and a cap. ($\div$ 3)

- Fungi have a mycorrhizal relationship with this organism.

  tree roots ($\div$ 4) leaves ($\times$ 2)

- Two organisms live and work together for the benefit of both organisms.

  mutualism (- original number) mycelium ($\times$ 3)

4. Write the final answers on the lines. _____ _____ _____

Lab 2.4.1A Pond Water Protists

QUESTION: Does pond water host a variety of protists?

HYPOTHESIS: _____

_____

EXPERIMENT:

You will need:	• 2 eyedroppers	• 2 coverslips	• resources on pond life
• pond water	• 2 microscope slides	• microscope	• cotton

Steps:
1. Allow the pond water to settle.
2. Predict whether you think you will find the same kinds of protists in the top and in the bottom of the sample. _____

   _____

3. Predict whether you think there will be more protists in the top or in the bottom sample of water. Record your prediction. _____

   _____

4. Using an eyedropper, take a drop of water from the top of the pond-water sample. Prepare a microscope slide. Lay a few strands of cotton across the drop of water to help slow the protists down. Carefully lower the coverslip over the water sample. Observe the slide through the microscope. Slowly move the slide around to look for different protists.
5. Sketch four different organisms that you see. Briefly describe each one's color, shape, movement pattern, structure, feeding pattern (if appropriate), and evidence that it responds to stimuli (such as light or obstacles).

A   **B**   **C**   **D**

_____   _____   _____   _____

_____   _____   _____   _____

_____   _____   _____   _____

_____   _____   _____   _____

Lab 2.4.1A Pond Water Protists

6. Use resources that you have available to identify each organism.

A. _____   C. _____

B. _____   D. _____

7. Using a clean eyedropper, take a drop of water from the bottom of the pond-water sample. Repeat Steps 4–6. Record your observations.

| E | F | G | H |

_____   _____   _____   _____

_____   _____   _____   _____

_____   _____   _____   _____

_____   _____   _____   _____

ANALYZE AND CONCLUDE:

1. Compare your hypothesis to the results. Did you observe a variety of protists? _____

_____

_____

2. How were the various organisms the same? _____

_____

_____

How were they different? _____

_____

3. Compare your predictions to the results. Did you find the same kinds of organisms in the top sample as in the bottom sample? _____

_____

_____

Why do you think this is so? _____

_____

Lab 2.4.2A Euglena

QUESTION: How do euglena respond to light?

HYPOTHESIS: _____

_____

EXPERIMENT:

You will need:	• concentrated euglena	• microscope slide	• microscope
• eyedropper	culture in a petri dish	• coverslip	• black paper

Steps:

1. Sketch a euglena. Include and label all of its features.

2. Prepare a microscope slide with a drop of the euglena culture. Carefully lower the coverslip over the specimen. Examine the slide under a microscope using low and high power.

3. Draw a euglena under low power and then high power. Label any features that you observe. Note its color.

Low Power

High Power

Color _____

Color _____

Lab 2.4.2A Euglena

4. Describe the euglena culture in the petri dish. _____

_____

5. Cover half of the petri dish with a small piece of black paper. Predict what you
   think will happen. _____

_____

6. After 20 minutes, uncover the dish. Record your observations. _____

_____

ANALYZE AND CONCLUDE:

1. How do euglena respond to light? _____

_____

2. Compare your hypothesis to how the euglena actually responded to light. _____

_____

3. Why do you think euglena respond this way? _____

_____

4. What structures of the euglena help it respond to light? _____

_____

5. Compare your sketch of a euglena to the drawings of your observations. What
   structures did you observe? _____

_____

Which structures were you not able to see? _____
Why do you think you were not able to observe those structures? _____

_____

6. Do you think that euglena are more likely to be found near the surface of the water
   or well below the surface? Why? _____

_____

7. Why do you think euglena are classified as plantlike protists? _____

_____

Lab 2.4.3A Stream Water Algae

QUESTION: What conditions are best for algae growth in stream water?

HYPOTHESIS: _____

_____

EXPERIMENT:

You will need:	• leaves from the bottom	• ammonia test kit
• 3 wide-mouthed jars	of a stream, not cleaned	• marking pencil
• water from a stream	• rock from stream, not cleaned	

Steps:

1. Fill one jar with just stream water. Label this jar with the word *Control*.
2. Fill a second jar with stream water; place a rock from the stream in the jar. Make sure not to clean off the rock. Label the jar with the word *Rock*.
3. Fill the third jar with stream water; add several leaves from the stream to the jar. Do not rinse or attempt to clean off the leaves. Mark the jar with the word *Leaves*.
4. Examining the contents of the three jars, predict which water will experience the

   most algae growth. _____

   _____

5. Place the jars in an area where they will receive indirect sunlight. The optimum temperature for algae growth is between 20°C–24°C. Temperatures above 35°C are lethal to algae.
6. Observe the changes in the jars over a period of two weeks. Record your observations.

Week 1	Observations
Day 1	
Day 2	
Day 3	
Day 4	
Day 5	
Day 6	
Day 7	
Week 2	**Observations**
Day 1	
Day 2	
Day 3	
Day 4	
Day 5	
Day 6	
Day 7	

Lab 2.4.3A Stream Water Algae

7. Test the ammonia content of each jar. Record this data.

Ammonia Content	
Control	
Rock	
Leaves	

ANALYZE AND CONCLUDE:

1. Which jar grew the most algae? _____

_____

2. Which jar grew the least algae? _____

_____

3. Compare your prediction in Exercise 4 above to the results. _____

_____

4. What is the relationship between the ammonia level and the rate of algae growth?

_____

_____

5. Knowing the results, rewrite your hypothesis using the information you have

learned about sunlight and ammonia. _____

_____

Lab 2.4.4A Ciliates

QUESTION: How do ciliates behave?

HYPOTHESIS: _____

_____

EXPERIMENT:

You will need:	• microscope	• coverslip	• eyedropper
• paramecium culture	• microscope slide	• green algae culture	

Steps:
1. Prepare a microscope slide of the paramecium culture.
2. Slowly move the slide around until you find a paramecium. Observe it under low and high power. You may have to continue to move the slide around to keep it in view since paramecium can move quickly. Reduce light by adjusting the diaphragm in order to see more details.
3. Sketch your observations.

Low Power   **High Power**

4. Remove the coverslip and add a small drop of green algae onto the paramecium culture. Replace the coverslip and watch the paramecium for 10 minutes under high power. What changes do you observe inside the paramecium? Record your

   observations. _____

_____

ANALYZE AND CONCLUDE:

1. Look at the labeled diagram of a paramecium in your textbook. List the organelles

that you were able to observe. _____

_____

Lab 2.4.4A Ciliates

2. Did you observe any organelles on high power that were not observable at low power? _____

_____

3. Describe how the paramecium moved through the water. _____

_____

_____

_____

4. What is the purpose of a paramecium's cilia? _____

_____

_____

5. Describe the process of the paramecium eating green algae. _____

_____

6. What other behaviors did you notice while observing the paramecium? _____

_____

_____

7. How do paramecia reproduce? _____

_____

8. Research the *Paramecium bursaria*. How is this particular species unique from other species of paramecium? _____

_____

_____

9. Compare your hypothesis to your observations. _____

_____

Name: _____   Date: _____

Lab 2.4.6A Slime Mold

QUESTION: Is slime mold a unicellular or multicellular organism?

HYPOTHESIS: _____

_____

EXPERIMENT:

You will need:	• coverslip	• small glass bowl	• water	• sharp needle
• slime mold	• microscope	that fits inside the	• cotton swab	• yeast
• toothpick	• paper towel	large bowl	• filter paper	• eyedropper
• microscope slide	• large bowl	• rubber band	• hand lens	

Steps:
1. Use a toothpick to prepare a microscope slide with a small amount of slime mold.
2. Examine the slime mold under high and low power. Sketch your observations.

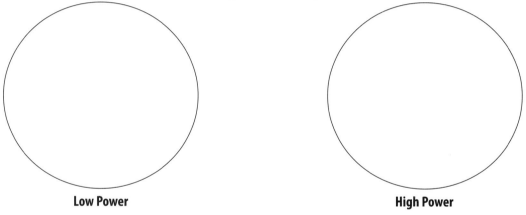

Low Power   **High Power**

3. Cover the small bowl with the paper towel. Wrap the rest of the paper towel underneath the bowl. Secure the paper towel with a rubber band.
4. Set the small bowl in the large bowl.
5. Add water to the large bowl until the water reaches three-fourths of the way up the large bowl. The water should wick up the paper towel.
6. Using a cotton swab, transfer a small amount of slime mold onto the filter paper.

7. Examine the slime mold with a hand lens. Record your observations. _____

_____

8. Use a sharp needle to poke a hole in one branch of the slime mold. Observe that

branch for a few minutes with a hand lens. Record your observations. _____

_____

9. Carefully place filter paper onto the paper towel that is covering the small bowl.
10. Sprinkle yeast on the paper towel next to the filter paper.
11. Add 3 drops of water to the yeast and 3 drops of water to the slime mold.

Lab 2.4.6A Slime Mold

12. Set the bowl in a cool, dark place. Observe it daily for three days. Record your observations.

	Observations
Day 1	
Day 2	
Day 3	

ANALYZE AND CONCLUDE:

1. What happened to the needle puncture in the slime mold? _____

_____

2. How did the slime mold change after three days? _____

_____

3. How is the slime mold like individual cells? _____

_____

4. How is the slime mold like one organism? _____

_____

5. Critique your hypothesis using the results of this experiment. _____

_____

6. Rewrite your hypothesis to make a true statement. _____

_____

_____

Name: _____ Date: _____

Protozoa

	Amoeba	Ciliate	Zooflagellate
How it eats			
How it moves			
How it reproduces			
Shape			
Other facts			

Protist Practice

```
X  M  F  N  O  A  I  J  P  F  C  Z  A  L  R  R  Z  W  E  W
S  W  I  E  O  C  U  S  R  H  B  Y  X  E  D  O  A  M  D  O
H  T  A  C  D  I  E  T  L  S  R  E  D  D  A  L  B  R  I  A
C  T  I  Q  R  U  T  O  O  D  M  T  K  G  X  S  T  G  T  K
Z  G  U  P  D  O  R  A  A  T  S  X  G  L  N  E  Z  G  D  R
G  V  Y  O  E  O  O  I  G  I  R  F  X  C  N  T  Q  Q  E  C
L  K  P  X  P  S  R  R  T  U  U  O  B  Q  F  Y  K  D  R  H
E  O  N  H  R  A  Z  C  G  Q  J  A  P  E  N  H  C  V  P  E
D  Z  Y  O  L  T  O  T  G  A  X  N  L  H  N  P  U  B  G  T
U  L  B  A  T  T  F  W  R  N  N  T  O  U  Q  O  R  R  D  E
L  L  M  V  O  K  P  K  S  D  Q  I  R  C  G  R  X  N  Y  R
B  Y  Y  R  E  T  N  E  S  Y  D  D  S  V  H  P  C  H  Q  O
F  M  P  R  O  I  R  A  D  Y  C  C  W  M  O  A  I  A  G  T
S  G  R  C  X  S  E  B  L  A  D  E  S  H  L  S  L  U  Z  R
Y  P  E  K  I  M  P  A  B  P  B  X  F  G  D  E  I  P  H  O
T  U  T  V  I  T  C  M  Q  L  I  Y  Z  F  F  R  A  J  S  P
M  T  Q  L  P  Q  T  S  I  T  O  R  P  O  A  O  P  P  G  H
P  M  W  V  A  V  H  Q  W  B  S  B  O  G  S  K  N  S  K  O
P  N  S  P  O  R  O  Z  O  A  N  S  H  D  T  D  A  J  G  C
J  W  J  E  E  L  X  A  L  R  I  K  L  G  S  D  A  E  X  E
```

Fill in the correct words and then find the words in the puzzle above.

1. A _____ is defined as not being an animal, plant, fungus, or bacterium.

2. A _____ is a tiny living thing that can be seen only under a microscope.

3. An _____ is an organism that makes its own food from simple substances.

4. A _____ is an organism that cannot photosynthesize, so it has to eat.

Protist Practice

5. Tiny organisms called _____ are found floating on or near the water's surface.

6. Sometimes dinoflagellates multiply so fast that they color a section of ocean water, forming a _____.

7. Plants and many protists use _____ to capture light for photosynthesis.

8. Brown algae depend on grapelike bulbs called _____ to float within range of sunlight.

9. Brown algae do not have leaves, but they have leaflike structures called

_____.

10. Brown algae do not have stems, but they have stemlike structures called

_____.

11. Brown algae do not have roots, but they have rootlike structures called

_____.

12. Amoebas use a false foot called a _____ to move around and to gather food.

13. If a person drinks water contaminated with certain amoebas, he or she might get

_____.

14. Ciliates use tiny hairlike projections called _____ to move themselves around.

15. Ciliates use a process called _____ to attach themselves to each other and exchange genetic information.

16. _____ cannot move on their own and cannot make their own food, so they survive as parasites.

17. _____, which causes muscle aches, fevers, and chills, is caused by sporozoans.

18. Some protists called _____ keep the food webs moving by eating dead or decaying materials.

Name: _____   Date: _____

Lab 3.1.3A Moss Structure

QUESTION: How is moss structured?

HYPOTHESIS: _____

_____

EXPERIMENT:

You will need:	• hand lens	• coverslip	• eyedropper
• moss plants	• paper	• forceps	• water
• toothpick	• microscope slide	• pencil with eraser	• microscope

Steps:

1. With the toothpick, isolate a single moss plant that does not have a stalk and capsule. Examine the plant with the hand lens and sketch it, labeling the rhizoids and the leaflike structures. On the diagram, mark where the zygote would form.

2. With the toothpick, isolate a single moss plant that has a stalk and capsule. Examine the plant with the hand lens and diagram it, labeling the rhizoids, leaflike structures, stalk, and capsule. Mark where the spores are located.

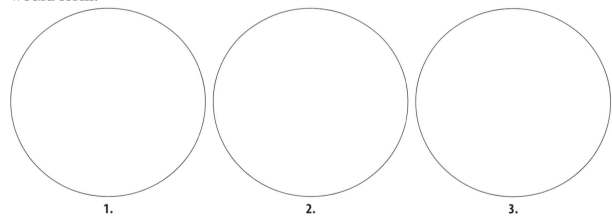

1.

2.

3.

3. Choose a brown or black spore capsule. Place it on a piece of paper and examine it with the hand lens. Sketch the structure, noting the lid on the top of the capsule. Place the capsule on the microscope slide. Carefully hold the capsule in place with the forceps. Then use a pencil eraser to press down on the capsule until it breaks and releases the spores.

4. Discard the capsule and add a drop of water to the spores. Cover the slide with the coverslip and examine the spores under low and high power of the microscope.

Lab 3.1.3A Moss Structure

5. Sketch four of the spores under low and high power.

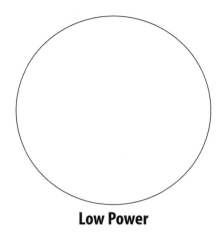

Low Power

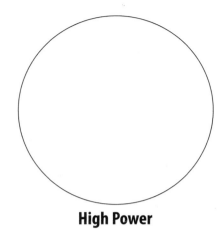

High Power

6. Choose a green capsule. Try to break it open as described in Step 3.

ANALYZE AND CONCLUDE:

1. Which part of the moss plant is the sporophyte? _____

_____

2. Which part of the moss plant is the gametophyte? _____

_____

3. What is the function of the lid on the capsule? _____

_____

4. What happened when you tried to break open the green capsule? Why do you think

this happened? _____

_____

5. Compare your hypothesis to your observations. What did you learn? _____

_____

Name: _____ Date: _____

Lab 3.1.4A Fern Structure

QUESTION: How is a fern structured?

HYPOTHESIS: _____

_____

EXPERIMENT:

You will need:	• 2 microscope slides and coverslips	• microscope
• mature fern plant with spore cases	• eyedropper	
• hand lens	• water	

Steps:
1. Examine the fern plant.
2. Sketch the plant. Label a frond, rhizome, sori (spore cases), and fiddlehead.

3. Scrape a few spore cases off the frond onto a microscope slide.
4. With an eyedropper, add water to the spore cases. Cover the spore cases with a coverslip and examine them under low and high power of the microscope.
5. Sketch the spore cases as they appear under low power. Remove additional spore cases from a frond and prepare another slide without water. Observe and sketch the spore cases under low power.

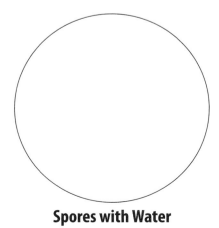

Spores with Water

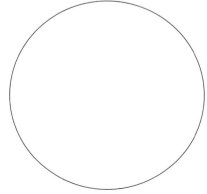

Spores without Water

Lab 3.1.4A Fern Structure

6. After 10 minutes, observe both sets of spore cases again.
7. Sketch the spore cases and spores.

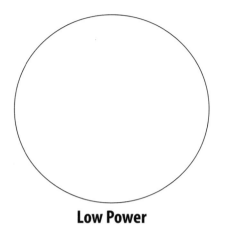

Low Power

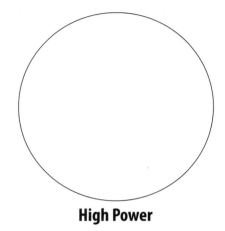

High Power

ANALYZE AND CONCLUDE:

1. How were the spores different when you observed them again after waiting 10 minutes? _____

_____

2. Using your observations, describe how a fern is structured. _____

_____

_____

3. How do your observations compare with your hypothesis? _____

_____

4. Explain how the structure of a fern supports that it was uniquely designed by God.

_____

_____

Lab 3.1.5A Root Comparison

QUESTION: How do taproots and fibrous roots compare?

HYPOTHESIS: _____

_____

EXPERIMENT:

You will need:	• 10 bean seeds	• 2 ziplock bags
• 10 grass seeds	• 2 damp paper towels	

Steps:
1. Place 10 grass seeds on a damp paper towel and seal them inside a ziplock bag. Place 10 bean seeds on the other damp paper towel and seal them inside a ziplock bag.
2. Place the bags in indirect sunlight for several days. Remoisten the paper towels if necessary.
3. Compare the growth of roots and shoots for both kinds of seeds for nine days. Note how many seed leaves you see.
4. Record and sketch your observations on **WS 3.1.5A Taproots and Fibrous Roots.**

ANALYZE AND CONCLUDE:

1. Which plant has fibrous roots? _____ Which has a taproot? _____

2. How many seed leaves did the grass seeds produce? _____ The bean seeds? _____

3. Is grass a monocot or dicot? _____ The beans? _____

4. Research to find out if there is a correlation between a plant being a monocot or a dicot and the type of root system it has. _____

_____

5. List the differences and similarities between the taproots and the fibrous roots.

_____

_____

6. What do you think are the advantages of each type of root? _____

_____

_____

_____

Lab 3.1.5B Taproot Structure

QUESTION: What is the structure of a taproot?

HYPOTHESIS: _____

_____

EXPERIMENT:

You will need:	• 2 carrots	• knife	• hand lens

Steps:
1. Carefully cut off the top of a carrot. Then, cut a cross section from the top of the carrot. Then cut the remaining carrot in half lengthwise.
2. Examine the slices of carrot with the hand lens. Draw a diagram of the inner and outer parts. Label the epidermis, root hairs, cortex, xylem cells, and phloem cells.

Lengthwise   **Cross section**

ANALYZE AND CONCLUDE:

1. How is a taproot structured? _____

_____

_____

2. List some other taproots that people use as foods. _____

_____

_____

3. Why are taproots used for food more often than fibrous roots? _____

_____

_____

Lab 3.1.5C Tree Ring Study

QUESTION: Are all the tree rings in a tree trunk the same size?

HYPOTHESIS: _____

_____

EXPERIMENT:

You will need:	• cross section of a tree trunk	• hand lens

Steps:

1. Sketch the cross section. Label the pith, xylem, phloem, bark, heartwood, cambium, and sapwood.

2. Examine each part of the cross section with the hand lens. Sketch a magnified section of each part.

3. Determine the age of the tree when it was cut by counting either the dark or light rings. How old was the tree? _____

4. Look for variations in the ring width. Describe any variations. _____

_____

5. Describe any knots. _____

_____

Lab 3.1.5C Tree Ring Study

6. Record any other interesting facts. _____

_____

ANALYZE AND CONCLUDE:

1. What are the tiny channels in the sapwood of the tree? _____

_____

2. Why did you count only the dark or light rings and not both? _____

_____

3. Were all the tree rings the same size? _____ What does this indicate?

_____

_____

4. How does your hypothesis compare to your observations? _____

_____

5. What would you do to ensure the results to this experiment were more accurate?

_____

_____

6. What might the knots indicate? _____

_____

7. Describe the life of the tree. For example, can you tell when the tree was competing with neighbors for light? Which years had less rainfall? Is there evidence of a fire?

_____

_____

_____

_____

Lab 3.1.5D Stomata

QUESTION: How do stomata work?

HYPOTHESIS: _____

_____

EXPERIMENT:

You will need:	• metric ruler	• microscope slide	• saltwater
• hand lens	• forceps	• coverslip	• paper towel
• lettuce leaf in water	• eyedropper	• microscope	

Steps:

1. Remove the lettuce leaf from the water and examine both sides of the leaf with the hand lens. Describe what you see. _____

   _____

2. Tear a 5 cm square from the lettuce leaf. Fold the square and remove a thin layer of the epidermis with the forceps.

3. Place a drop of water on the microscope slide and place the epidermis in the drop. Cover it with the coverslip.

4. Observe the epidermis under low power with the microscope. Sketch the epidermis. Label an epidermal cell, a guard cell, a stoma, and a chloroplast.

Low Power

5. Place a drop of saltwater on the slide, touching the edge of the coverslip. Touch a small piece of paper towel to the opposite edge of the coverslip so that the saltwater flows underneath the slip to the paper towel. Wait about five minutes.

Lab 3.1.5D Stomata

6. Repeat Step 4. Sketch your observation and label your drawing.

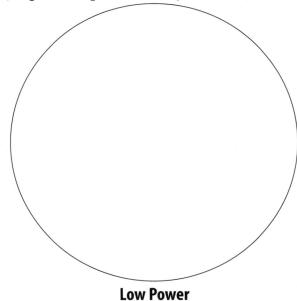

Low Power

ANALYZE AND CONCLUDE:

1. How are guard cells different from other epidermal cells? _____

_____

2. Compare the guard cells before and after you added the saltwater. _____

_____

3. How do you think the guard cells in a dry plant would look? _____

_____

4. How does this design protect the plant? _____

_____

5. Check your hypothesis. Make any necessary revisions and rewrite it to include

what you have learned from this experiment. _____

_____

_____

_____

Plant Search

Find 10 plants indoors or outdoors. Use field guides or other resources to identify them. Classify each plant as nonvascular or vascular. Further classify the vascular plants as gymnosperms or angiosperms and according to their life spans (annuals, biennials, or perennials).

Common Name	Latin Name	Vascular or Nonvascular	Gymnosperm or Angiosperm	Life Span

Plant the Words   **WS 3.1.1B**

Match each term to its definition and give an example for each.

_____**1.** angiosperm

_____**2.** annual

_____**3.** biennial

_____**4.** gymnosperm

_____**5.** perennial

_____**6.** gametophyte

_____**7.** sporophyte

_____**8.** nonvascular plant

_____**9.** vascular plant

a. a type of plant that does not have vessels to transport materials

b. the gamete-producing phase of a plant

c. a plant that completes its life cycle in one growing season

d. a type of plant that has vessels to transport materials

e. a flowering plant that lives for more than two years

f. a flowering plant that produces seeds

g. a plant that produces seeds but no flowers

h. a plant that requires two growing seasons to complete its life cycle

i. the spore-producing phase of a plant

Moss Life Cycle

Number the squares in the life cycle order of moss.

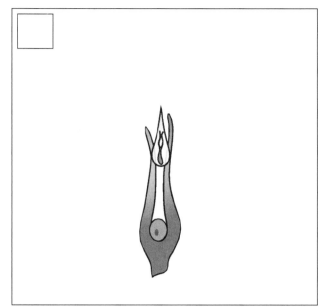

Vascular Seedless Plants

Plants and Their Features	Similarities to Plants with Seeds	Differences from Plants with Seeds
Ferns		
Leaflike structures		
Stemlike structures		
Rootlike structures		
Horsetails		
Leaflike structures		
Stemlike structures		
Rootlike structures		
Club Mosses		
Leaflike structures		
Stemlike structures		
Rootlike structures		

Taproots and Fibrous Roots

	Grass Seeds	Number Sprouted	Seed Leaves	Sketch
Day 1				
Day 2				
Day 3				
Day 4				
Day 5				
Day 6				
Day 7				
Day 8				
Day 9				

	Bean Seeds	Number Sprouted	Seed Leaves	Sketch
Day 1				
Day 2				
Day 3				
Day 4				
Day 5				
Day 6				
Day 7				
Day 8				
Day 9				

Tree Rings

Examine the tree rings and answer the following questions:

1. What makes the rings? _____

_____

2. How old was the tree? _____

_____

3. Where is the oldest part of the tree? _____

_____

How do you know? _____

_____

4. Describe what might have caused some of the differences in the rings. _____

_____

5. Imagine that a growing tree engulfed some of the wire of a fence as it grew. In the coming years, what would happen to the height of this part of the fence? Why?

_____

_____

Seed Plants   **WS 3.1.5C**

Complete the concept map by providing the three main parts of seed plants and the functions of each part.

Concept Map

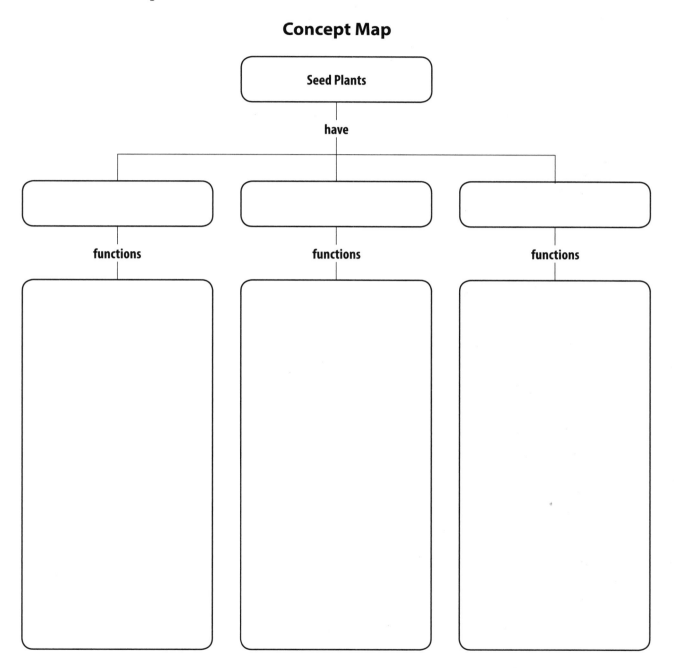

Lab 3.2.1A Photosynthesis

QUESTION: How does the lack of sun affect photosynthesis?

HYPOTHESIS: _____

_____

EXPERIMENT:

You will need:	• house plant that needs lots of light	• dark paper or black electrical tape

Steps:

1. Observe the leaves and write down your observations about their shape and color. _____

2. Cover part of several leaves with the paper or tape. Be careful not to pierce the leaves.

3. Predict how the partially covered leaves will be affected after five days. _____

_____

4. Place the plant in a sunny place or under a plant light for at least five days.

5. Remove the paper or tape and observe the change in the leaves. Record your observations. _____

ANALYZE AND CONCLUDE:

1. In what part of the plant does photosynthesis occur? _____

2. What substance is responsible for photosynthesis? What color is this substance?

_____

3. What would happen to the plant if you covered all its leaves? _____

_____

4. Where have you seen effects like this in nature? _____

_____

5. Compare your prediction in Step 3 above to the results. _____

_____

6. Did the results of this experiment support your hypothesis? If not, rewrite your hypothesis. _____

Lab 3.2.1B Plants and Pollution

QUESTION: How does pollution affect the life processes of plants?

HYPOTHESIS: _____

_____

EXPERIMENT:

You will need:	• water	• black plastic
• 2 large jars	• 4 elodea plants	• 2 plant lights
• gravel	• black construction paper	• used motor oil
• metric ruler	• tape	

Steps:
1. Put 5 cm of gravel in the bottom of each jar. Fill the jars with water and let them stand for 24 hours.
2. Plant two elodea plants in each jar. Let the jars stand for 24 hours. Describe the

    elodea plants and the water. _____

    _____
3. Cover the sides and bottoms of the jars with black plastic and then with black construction paper.
4. Set a plant light above each jar. Leave the lights on for 13 hours each day throughout the rest of the experiment.
5. Let the jars stand for 24 hours.
6. Carefully pour some used motor oil into one of the jars until the oil covers the surface of the water. Let the jars stand for one week.
7. Remove the plastic and construction paper and observe the results. Record your

    observations. _____
8. Give the oil to your teacher to dispose of responsibly (as a hazardous material).

ANALYZE AND CONCLUDE:

1. Why were the plants in the jar with oil unable to perform photosynthesis?

_____

2. How can you tell that the plant in the jar without oil performed photosynthesis?

_____

3. Compare your hypothesis to the results. _____

_____

4. How could an oil spill affect aquatic life? _____

_____

Lab 3.2.2A Flower Structure

QUESTION: How is a flower structured for reproduction?

HYPOTHESIS: _____

_____

EXPERIMENT:

You will need:	• hand lens	• microscope slide	• microscope
• flower	• eyedropper	• coverslip	
• forceps	• water	• scalpel	

Steps:
1. Observe the flower carefully. Sketch it.
2. Count the sepals and record the number found on the data chart. Label the sepals on your sketch.
3. Observe the petals. Remove them one at a time with the forceps. Record the number of petals on the data chart. Label the petals on your sketch.
4. Observe the stamens. Record the number of stamens on the data chart. Label the stamens on your sketch. Observe one under a hand lens. Sketch the magnified stamen. Label the anther and filament.
5. Locate the pistil(s). Record the number of pistils on the data chart. Observe one under a hand lens. Sketch the magnified pistil. Label the stigma, style, and ovary.

Number of Flower Parts

Sepals	Petals	Stamens	Pistils

6. Place a drop of water on a microscope slide. Remove an anther with the forceps and place it on the slide. Carefully cut the anther open with the scalpel. (Caution: Cut away from your body on a flat, firm surface.)
7. Use the forceps to gently shake the anther so that the pollen grains fall into the water on the slide. Remove the anther and cover the pollen with a coverslip.

Lab 3.2.2A Flower Structure

8. Examine the pollen under low power and then under high power. Sketch one grain of pollen under low power and one under high power.

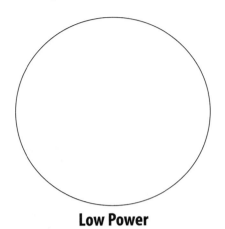

Low Power

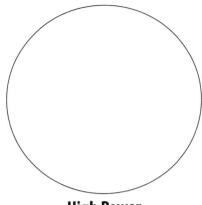

High Power

9. With the forceps, remove the pistil. With the scalpel, cut open the ovary at the base. Locate the ovules. Examine them with the hand lens. Sketch the magnified ovary and ovules.

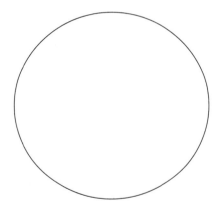

ANALYZE AND CONCLUDE:

1. Was the flower a monocot or a dicot? Explain how you know this. _____

_____

2. What are the female parts of the flower? _____

3. What are the male parts of the flower? _____

4. Which parts are neither female nor male? _____

5. Compare your hypothesis to your observations. Rewrite your hypothesis if

necessary. _____

_____

Name: _____ Date: _____

Lab 3.2.3A Pine Cone Structure

QUESTION: How is a pine cone structured?

HYPOTHESIS: _____

_____

EXPERIMENT:

You will need:	• pine cone samples from 3 different	• water
• paper towels	species	

Steps:

1. Label the boxes below with the name of the tree each cone came from. Describe the seeds, scales or other coverings, and any other special features of each cone.

Cone 1 _____ **Cone 2** _____ **Cone 3** _____

2. Determine where in the life cycle the cone is. Is the cone ready to open and disperse its seeds?

Cone 1 _____

Cone 2 _____

Cone 3 _____

Lab 3.2.3A Pine Cone Structure

   3. Dip the cones in water. How do they respond?

      Cone 1 _____

      Cone 2 _____

      Cone 3 _____

   4. Lay the cones on paper towels to dry out for a day. How do they respond?

      Cone 1 _____

      Cone 2 _____

      Cone 3 _____

   5. Look for seeds inside the cones. If the seeds are gone, look for the impression of the seedbed. Usually there are two seeds at the base of each scale.

ANALYZE AND CONCLUDE:

1. Why do you think that cones are designed to respond as they do when they get wet?

_____

_____

2. Why do you think that the cones are designed to respond as they do when they are dry? _____

_____

3. Did your observations support your hypothesis? If yes, explain why. If not, rewrite your hypothesis to reflect your observations. _____

_____

Lab 3.2.4A Plants and Light

QUESTION: How do plants respond to different lights?

HYPOTHESIS: _____

_____

EXPERIMENT:

You will need:	• 2 pieces each of red, blue, and green transparent film	• 6 boxes to hold the plants
• 6 plants of the same species		

Steps:

1. Place each plant in a box.
2. Place one plant outside in bright sunlight and another plant in shade.
3. Cover the openings of three of the remaining boxes with transparent film of a different color. Cover the opening of the last box with transparent film of all three colors. Place the plants in direct sunlight, either outside or inside the classroom.

4. Predict how well each plant will grow. Explain your predictions. _____

_____

5. Observe the plants for two weeks and create a data chart to record your observations.

ANALYZE AND CONCLUDE:

1. Compare the shades of green in the plants grown in the sun and in the shade.

Explain the results. _____

_____

_____

_____

2. Compare the growth of the plants grown under the red, blue, and green

transparencies. _____

_____

_____

_____

3. Compare your hypothesis and predictions to the results. _____

_____

Lab 3.2.4B Root Growth

QUESTION: Which way will roots grow?

HYPOTHESIS: _____

_____

EXPERIMENT:

You will need:	• cotton	• cardboard	• tape
• 3 bean seedlings	• water	• 3 test tubes	• eyedropper

Steps:
1. Carefully wrap cotton around the roots and the lower shoot of the three bean seedlings, taking care not to damage the plant. Dip each seedling into water.
2. Place each seedling into the mouth of a test tube. Add more cotton as necessary to position the seedling snugly.
3. Tape each test tube to the cardboard—one facing sideways (label this *A*), one facing up (label this *B*), and one facing down (label this *C*).

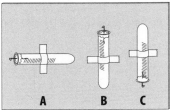

Prop the cardboard on a shelf or tack it to a bulletin board where it will not be disturbed.

4. Predict how the stems and roots will grow in each test tube. _____

_____

5. Moisten the cotton daily with an eyedropper.
6. Observe the seedlings for a week. Create your own data chart and sketch the three seedlings each day on the chart.

ANALYZE AND CONCLUDE:

1. Which direction did the stems and the roots grow in each test tube? _____

_____

2. Did the results match your predictions? Explain. _____

3. To which stimuli did the stems respond? Were these positive or negative tropisms?

_____

4. To which stimulus did the roots respond? Was this a positive or negative tropism?

_____

Scatter the Seeds

Use this worksheet to complete *Try This: Scatter the Seeds*. Examine the seeds to determine whether they are structured to be dispersed by wind, animals, bursting, or water.

Seed	Drawing	Wind	Animals	Bursting	Water

Flower and Fruit Observation WS 3.2.2B

Write the name of the flower or fruit at the top of each column. Draw or describe each structure that you can observe in the appropriate column.

Anther					
Filament					
Ovary					
Pistil					
Seed					
Stamen					
Stigma					
Style					

1. How are stamens structured to pollinate flowers? _____

_____

2. How are the parts of the pistil structured to capture pollen? _____

_____

3. Where are the seeds found? _____

_____

Flower Anatomy

Label the parts of the flower. Describe the function of each part.

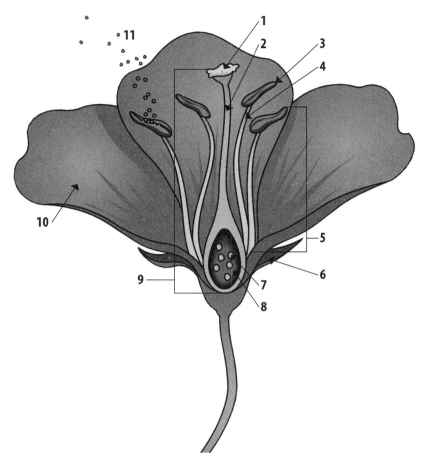

1. _____

2. _____

3. _____

4. _____

5. _____

6. _____

7. _____

8. _____

9. _____

10. _____

11. _____

Conifer Identification Key   WS 3.2.3A

1. a. Leaves are needlelike.. go to 2.
   b. Leaves are flattened and scalelike...................................... go to 11.

2. a. Leaves are clustered.. go to 3.
   b. Leaves are not clustered .... go to 6.

3. a. Clusters have two to five needles .... go to 4.
   b. Clusters have over 10 needles .... go to 5.

4. a. Clusters have five needles.. white pine.
   b. Clusters have two needles... Scots pine.

5. a. The needles are soft.. larch.
   b. The needles are stiff .... true cedar.

6. a. The twigs have pegs (leaf scars) .... go to 7.
   b. The twigs do not have pegs (leaf scars) .... go to 8.

7. a. The pegs are square and the needles are sharp.................. spruce.
   b. The pegs are rounded and the needles are blunt................ hemlock.

8. a. The buds are large and pointed .... Douglas fir.
   b. The buds are not large and pointed..................................... go to 9.

9. a. The terminal buds are round and clustered........................ true fir.
   b. The terminal buds are not round and clustered .... go to 10.

10. a. The needles are green underneath .... yew.
    b. The needles are white underneath .... go to 11.

11. a. The needles are pointed .... redwood.
    b. The needles are blunt... hemlock.

12. a. All leaves are short... Giant Sequoia.
    b. Some leaves are not sharp .... go to 13.

13. a. The cones are round .... go to 14.
    b. The cones are not round .... go to 15.

14. a. The cones are soft and leathery .... juniper.
    b. The cones are woody.. go to 15.

15. a. The cones are shaped like rose buds .... red cedar.
    b. The cones are shaped like a duck's bill............................... incense cedar.

Conifer Identification Key

_____

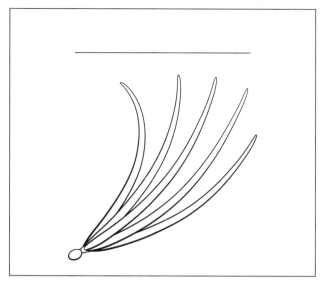

_____

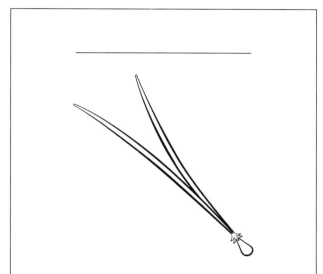

_____

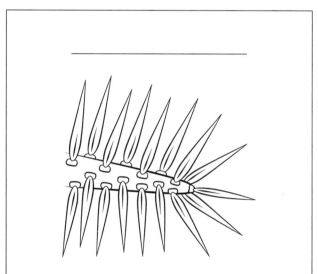

_____

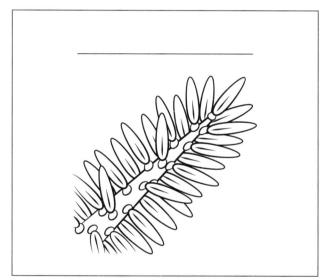

_____

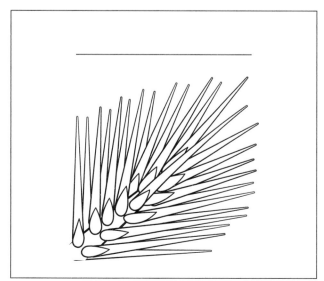

_____

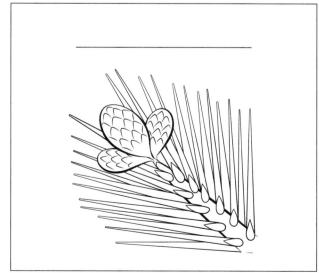

Plant Behavior

Describe or draw an example of the plant behavior listed below and explain how each helps the plant flourish.

Phototropism	**Gravitropism**

Thigmotropism	**Circadian Rhythms**

Lab 4.1.3A Hydra Reactions

QUESTION: How will a hydra react to various external stimuli (being fed; touched; adding vinegar, fertilizer, saltwater, and vegetable oil to the water)?

HYPOTHESIS: _____

_____

_____

EXPERIMENT:

You will need:	• water from an aquarium	• toothpick	• aquarium fertilizer
• hydra culture	• hand lens	• 4 eyedroppers	• saltwater
• petri dish	• fish food	• vinegar	• vegetable oil

Steps:

1. Remove a hydra from the culture and place it in the petri dish containing water from an aquarium.
2. Examine the hydra with a hand lens.
3. Sketch the hydra and label its tentacles, central cavity, and mouth.

4. Sprinkle a small amount of fish food into the petri dish. With a toothpick, gently poke a tiny bit of fish food near the hydra. Record your observations. _____

_____

5. Gently touch the hydra with the toothpick. Record your observations. _____

_____

6. Add a drop of vinegar (an acid) to the water. Record the hydra's response. _____

_____

7. Rinse the petri dish and repeat Steps 1 and 6 with a new hydra, using fertilizer instead of vinegar. What happened? _____

_____

Lab 4.1.3A Hydra Reactions

8. Rinse the petri dish and repeat Steps 1 and 6 with a new hydra, using saltwater instead of vinegar. Record your observations. _____ _____

9. Rinse the petri dish and repeat Steps 1 and 6 with a new hydra, using vegetable oil instead of vinegar. _____ _____

ANALYZE AND CONCLUDE:

1. Describe the best living conditions for a hydra. _____ _____ _____

2. Compare your hypothesis to the results of each stimuli. Did the hydra react the way you anticipated? _____ _____ _____ _____

3. Do you think it matters much if pollution harms creatures as tiny as a hydra? Why? _____ _____ _____

Lab 4.1.4A Worm Activity

QUESTION: How does temperature affect a worm's activity?

HYPOTHESIS: _____

_____

EXPERIMENT:

You will need:	• water	• access to a refrigerator and sunlight
• 2 similarly sized earthworms	• 2 plates	
• 2 paper towels	• stopwatch	

Steps:

1. Draw one of the worms and label the mouth, anus, clitellum, and setae.

2. Fold each paper towel into fourths and soak them in water. Let the excess water drip from the paper towels and then lay them each on a plate. Place a worm on each paper towel.

3. Find the large blood vessels on the top of each worm's body. Count the number of pulses per minute. Record the data.

   Earthworm A _____ Earthworm B _____

4. Put the plate and paper towel with Earthworm A in the refrigerator. Place the plate and paper towel with Earthworm B on a shelf in direct sunlight. Let them sit for one hour.

5. After one hour, take the worms out of the refrigerator and sunlight. Find the large blood vessels again and count the number of pulses per minute. Record the data.

   Earthworm A _____ Earthworm B _____

6. Observe each worm's activity. Gently poke them. Record your observations.

   Earthworm A _____

   Earthworm B _____

Lab 4.1.4A Worm Activity

ANALYZE AND CONCLUDE:

1. Which worm has the faster pulse rate? _____

2. Which worm is more active? _____

3. How does temperature affect a worm's activity? _____

_____

4. Compare your hypothesis to the results. _____

_____

5. Rewrite your hypothesis (if necessary) to reflect the results and to make a true

statement. _____

_____

Lab 4.1.4B Stimulus and Response

QUESTION: How does an earthworm respond to external stimuli (light, rubbing alcohol, and noise)?

HYPOTHESIS: _____

_____

EXPERIMENT:

You will need:	• box	• flashlight
• earthworm farm (or box filled with soil and earthworms)	• dark construction paper	• cotton swab
	• tape	• 70% rubbing alcohol
• damp newspaper	• scissors	• wind-up alarm clock

Steps:

1. Place an earthworm on a damp newspaper, and cover the newspaper with a box so that it is in complete darkness for one hour.

2. Make a cone with a piece of dark construction paper and tape and snip off the end of the cone.

3. Darken the room. Place the cone over a flashlight to create a small beam.

4. Predict what will happen when you shine the light on the earthworm. _____

_____

5. Remove the cover from the box and shine the small beam on the head end of the earthworm. Record your observations. _____

_____

6. Shine the beam on other parts (clitellum and anus) of the earthworm. Record your observations. _____

_____

7. Turn the room lights back on, choose a different worm, and place it on newspaper.

8. Dip a cotton swab in rubbing alcohol. Predict what will happen when you put the swab with rubbing alcohol near the worm's head. _____

_____

Lab 4.1.4B Stimulus and Response

9. Hold the swab with rubbing alcohol next to the worm's head. (Do not touch the worm with the swab.) What did the worm do? Record your observations. _____

_____

10. Hold the swab near the clitellum and anus of the worm. Record your observations. _____

_____

11. Predict what the worms in the earthworm farm will do when you let an alarm clock ring near them. _____

_____

12. Hold the face of the alarm clock against the worm farm. Let the alarm clock ring. Record your observations. _____

_____

ANALYZE AND CONCLUDE:

1. Did the earthworm respond to light? _____ How does this help the earthworm? _____

2. How accurate was your prediction? _____

3. Explain the earthworm's response to the smell of rubbing alcohol. _____

_____

_____

_____

4. How did your prediction compare to the worm's behavior in Step 8? _____

_____

5. Explain the earthworm's response to sound. _____

_____

6. Was your prediction accurate? Explain. _____

_____

7. Write three things you learned from this experiment. _____

_____

Animal Expressions

Many figures of speech involve animals. Write down as many as you can and decide if they have any scientific basis.

Expression	Meaning	What, if anything, is the scientific basis?

Spongy Stuff

Research to find the answers. Fill in the blanks.

1. Most sponges are _____ animals, but some are freshwater.

2. Sponges get food by _____.

3. Collar cells have little hairlike structures called _____.

4. _____ cover the sponge's outer surface and respond to touch.

5. Sponges have the ability to grow back missing parts through _____.

6. _____ cells maintain a flow of water throughout the sponge.

7. Sponges have a large opening called an _____.

8. Sponges have pores called _____.

9. _____ link together to form a type of skeleton for the sponge.

10. _____ form a jellylike layer between the collar cells and the epithelial cells.

11. Draw a sponge and label the osculum, ostium, spicule, collar cell, flagellum, amoebalike cell, and epithelial cell.

Mollusk Survey

1. Take a survey to determine the mollusks sold in a grocery store.

_____

_____

_____

_____

_____

_____

_____

_____

2. Choose one type of mollusk to research. Write facts about the mollusk including its size, where it is found, what it eats, and any unusual characteristics it has.

_____

_____

_____

_____

_____

_____

3. If possible, contact one of the distributors (with a toll-free number, e-mail, or web address) to determine how the mollusks were harvested. Explain whether you think they were harvested in a responsible manner.

_____

_____

_____

_____

_____

Spineless Language   WS 4.1.6A

Write the answer.

1. An animal without a backbone _____

2. Reproduction involving two parents _____

3. Reproduction involving one parent _____

4. A group of cells that performs a function _____

5. Two or more tissues working together _____

6. The regrowth of lost body parts _____

7. The cell with flagella in a sponge _____

8. The needlelike skeletal material in sponges _____

9. The stinging cell of cnidarians _____

10. A vase-shaped cnidarian _____

11. A bowl-shaped cnidarian _____

12. A snail uses this to scrape food _____

13. A growth organ that lines a mollusk's shell _____

14. A mollusk that has two hinged shells _____

15. A mollusk that has a single shell or no shell _____

16. An ocean mollusk that has tentacles _____

17. A skeleton on the inside of the body _____

18. A bulblike suction structure on echinoderms _____

Lab 4.2.2A Crayfish

QUESTION: How does a crayfish eat?

HYPOTHESIS: _____

_____

EXPERIMENT:

You will need:	• live crayfish	• small piece of fish
• aquarium	• hand lens	• pointer

Steps:

1. Using a hand lens, observe the live crayfish. Sketch and label the following features: *cephalothorax, carapace, legs, abdomen, antennae, swimmerets,* and *maxillipeds.*

2. Place a small piece of fish into the aquarium. Observe how the crayfish takes the food and eats. Record your observations. _____

_____

3. With a pointer, gently touch the crayfish on its carapace or antennae. Be careful not to injure it. Record your observations. _____

_____

Lab 4.2.2A Crayfish

ANALYZE AND CONCLUDE:

1. How does the crayfish take food and eat? _____

_____

2. How does the process by which the crayfish eats compare to your hypothesis?

_____

3. How does the crayfish move about when it is not threatened? _____

_____

4. How does the crayfish move about when it is threatened? _____

_____

5. Research to find five interesting facts about crayfish.

    a) _____

    b) _____

    c) _____

    d) _____

    e) _____

Lab 4.2.5A Grasshopper

QUESTION: Why is a grasshopper classified as an insect?

HYPOTHESIS: _____

_____

EXPERIMENT:

You will need:	• square of cheesecloth or netting	• hand lens
• large jar	• dirt and grass	
• live grasshopper	• rubber band	

Steps:

1. Pour a small amount of dirt in the bottom of the jar. Sprinkle some grass on top of the dirt.
2. Place the grasshopper in the jar. Cover the opening with the netting or cheesecloth. Secure it with the rubber band.
3. Observe and sketch the basic parts of the grasshopper: head, antennae, eyes, mouth, wings, thorax, legs, abdomen, and the spiracles (small openings on each section of the abdomen).
4. Use the hand lens to determine if the grasshopper is a male or a female. Females have a divided segment at the end of their abdomens for laying eggs. The male's abdomen has a blunt, rounded end.

Male

Female

Lab 4.2.5A Grasshopper

ANALYZE AND CONCLUDE:

1. How many body segments does the grasshopper have? _____

2. How many pairs of antennae does the grasshopper have? _____

3. Describe the grasshopper's mouth. _____

_____

4. How many legs does the grasshopper have? _____

5. Which pair of legs does the grasshopper use for jumping? How can you tell?

_____

_____

6. Which body part are the wings attached to? _____

7. Describe the two pairs of wings. How might they be used? _____

_____

_____

8. What are the four main characteristics of insects? _____

_____

_____

9. Defend why a grasshopper is classified as an insect. _____

_____

_____

Armor, Anyone?

List the advantages and disadvantages of having an exoskeleton.

Advantages	Disadvantages

Examine and describe various crustacean shells and specimens. Sketch and record the similarities and differences.

Crustacean	Similarities	Differences	Sketch

Examine and describe various crustacean shells and specimens. Sketch and record the similarities and differences.

Crustacean	Similarities	Differences	Sketch

Insect Observations

Observe five of the insects that you collected. Record your observations.

Insect	Sketch	Appearance	Activity

Speaking of Arthropods

Speaking of Arthropods continued   WS 4.2.5B

Across

5. The front two segments of some arthropods are fused to form a _____.

8. Lobsters, centipedes, ticks, and bees are all _____.

9. An insect's compound eye is formed by smaller parts called _____.

13. Arthropods with eight legs and two other pairs of appendages are called _____.

14. An insect's head is formed from hard plates called _____.

16. A many-footed arthropod with a flat body and one pair of legs per body segment is a _____.

Down

1. A crab has certain appendages that help it eat called _____.

2. When a grasshopper eats grass, it uses mouth parts called _____.

3. Shrimp, crabs, and wood lice all have hard shells; they are _____.

4. A scientist who studies insects is called an _____.

6. Arthropods have a hard outer covering called an _____.

7. An insect egg is transformed into an adult insect through _____.

8. The longest part of an insect is the _____.

10. Ticks can spread a bacterial infection called _____ _____.

11. A many-footed arthropod with a round body and two pairs of legs per body segment is a _____.

12. Insects breathe through a tube called a _____.

15. An insect's legs and wings grow from its center section, the _____.

Lab 4.3.2A Ectotherm Temperature

QUESTION: How does temperature affect ectotherms?

HYPOTHESIS: _____

_____

EXPERIMENT:

You will need:	• live goldfish	• crushed ice
• water	• large bowl	• warm water
• wide-mouthed jar	• thermometer	• stopwatch

Steps:

1. Place a small amount of water in the jar and place the goldfish in the jar. Make sure that the fish is barely covered by the water.
2. Place the jar inside a large bowl filled with water.
3. Place a thermometer in the jar near the glass where it can be easily read.
4. Fish breathe through their gills, but their mouth movements indicate their respiration rate. Observe the movements of the goldfish's mouth. Describe the

   movements. _____

   _____

5. Count the number of times the goldfish moves its mouth in 60 seconds. Record

   that respiration rate. _____

6. Add crushed ice very slowly to the water in the bowl (not the jar) until the water temperature is reduced to about 3°C. Avoid exciting the fish to minimize any

   shock factor. Predict what will happen to the goldfish's respiration rate. _____

   _____

7. Count the fish's mouth movements for another 60 seconds. Record the respiration

   rate. _____

8. Slowly replace the cold water in the bowl with warm water until the water temperature inside the jar reaches 30°C. Predict what will happen to the

   respiration rate. _____

   _____

9. Count the fish's mouth movements for another 60-second interval. Record the

   respiration rate. _____

Lab 4.3.2A Ectotherm Temperature

ANALYZE AND CONCLUDE:

1. Graph the water temperature and respiration rate. Give the graph a title and label the *x*-axis and *y*-axis.

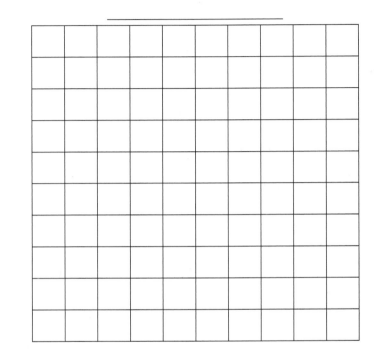

2. Explain how the temperature changes affected the fish. _____

_____

_____

3. Compare the results to your hypothesis and predictions. _____

_____

_____

_____

4. How else might temperature changes affect the fish? _____

_____

_____

5. How could you test these hypotheses? _____

_____

_____

Lab 4.3.5A Swim Bladder

QUESTION: How does a fish's swim bladder work?

HYPOTHESIS: _____

_____

EXPERIMENT:

You will need:	• water
• 2 L plastic soda bottle with cap	• eyedropper

Steps:
1. Fill the bottle to the very top with water.
2. Place the eyedropper in the bottle so that it floats just below the surface of the water in the bottle. (You may need to draw in or squeeze out a little water from the eyedropper.)
3. Screw the cap onto the bottle tightly so that no air or water leaks out when you squeeze the bottle.
4. Squeeze the bottle. Observe what happens.
5. Release the bottle. Observe what happens.

ANALYZE AND CONCLUDE:

1. What happens to the eyedropper when you squeeze the bottle? Why? _____

_____

_____

_____

2. What happens to the eyedropper when you release the bottle? Why? _____

_____

_____

3. How does increasing the eyedropper's density affect its ability to float? _____

_____

_____

4. What is the relationship between the density of a fish's swim bladder and the fish's

ability to float? _____

_____

_____

Lab 4.3.5A Swim Bladder

5. Draw a line graph to show the relationship between a swim bladder's density and the fish's ability to float.

6. Why does it help to take a big breath and hold it when you are floating on water?

_____

_____

_____

7. Compare your hypothesis to the results. _____

_____

_____

_____

_____

8. Explain how a fish's swim bladder works. _____

_____

_____

_____

_____

Name: _____   Date: _____

Lab 4.3.7A Feathers

QUESTION: How do feathers protect a bird?

HYPOTHESIS: _____

_____

EXPERIMENT:

You will need:	• contour feather	• stopwatch
• hand lens	• jar of water	• jar of oil

Steps:

1. Examine the feather with the hand lens. Sketch the feather and record your observations.

    _____

    _____

    _____

2. Dip the feather into the water for 60 seconds and then examine it with the hand lens. Sketch the feather and record your observations.

    _____

    _____

    _____

Lab 4.3.7A Feathers

3. Dip the feather into the oil for 60 seconds and then examine it with the hand lens. Sketch the feather and record your observations.

4. Dip the feather back into the water for 60 seconds and then examine it with the hand lens. Sketch the feather and record your observations.

ANALYZE AND CONCLUDE:

1. How did the feather react to the water the first time you dipped it? _____

_____

2. How did the feather react to the water the second time you dipped it? _____

_____

3. God gave birds feathers to insulate them and to help them fly. How do you think an

oil spill affects a bird's insulation and flying capabilities? _____

_____

_____

_____

Endotherm and Ectotherm

Compare a mammal (an endotherm) with a reptile (an ectotherm).

Endotherm	Mass	Behavior	Food type	Food mass	Food consumption	Activity

Ectotherm	Mass	Behavior	Food type	Food mass	Food consumption	Activity

Note: Food consumption is the food consumed per body mass per unit of time.

What do your observations of these animals tell you about the different metabolic needs of endotherms and

ectotherms? _____

_____

Tadpole Growth

Observe how tadpoles change. Note changes in the length of their tails, in their back leg development, in their foreleg development, in the widening of their mouths, in the enlargement of their eyes, and in their body shape. Note behavioral changes in the tadpoles as they grow as well.

Observations and Sketches

1. What is the initial length of the caudal fin (tail)? _____

_____

2. Describe the tadpole's mouth initially. _____

_____

3. What happens to the mouth as the tadpole matures? _____

_____

4. Which legs develop first? _____

_____

5. Describe the development of the tadpole. _____

_____

_____

_____

_____

Sweating

EXPERIMENT:

You will need:	• thermometer	• cotton ball	• 70% rubbing alcohol

1. Leave a thermometer on the table for 3 minutes to determine the room's temperature. Record the temperature.
2. Moisten a cotton ball with rubbing alcohol and spread a thin layer of cotton around the thermometer's bulb. Blow on the cotton 20 times. Record the temperature.

Initial Temperature	**Final Temperature**

ANALYZE AND CONCLUDE:

1. What happened to the temperature? Why? _____

_____

_____

2. Relate the results of this experiment to the sweating process in endotherms. How

can sweating help regulate the body temperature of endotherms? _____

_____

_____

_____

_____

Metabolic Rate Graph

Make a graph comparing the weights and metabolic rates of various mammals. Include a title for your graph. Label the *x*-axis *Relative Mass* and the *y*-axis *Relative Metabolic Rates*. Plot points for the following mammals using their approximate mass and their approximate metabolic rate compared with that of humans. (You could also research the metabolic ranges of larger animals and add those to your graph.)

- Horse: 650 kg (metabolic rate one-half that of a human)
- Human: 80 kg
- Cat: 5 kg (metabolic rate 2 times that of a human)
- House mouse: 0.03 kg (metabolic rate 5 times that of a human)
- Pygmy white-toothed shrew: 0.002 kg (metabolic rate 30 times that of a human)

Brain Stretchers

Use your reasoning ability to try to solve the puzzles.

1. Ashley, Brigón, Casey, and Devon have the following hobbies: art, basketball, cooking, and playing the drums. Using the facts below, figure out which student has which hobby.

- The name of at least one student begins with the same letter as that student's hobby.
- The artist's name does not contain an *a*.
- The basketball player and the drummer have the same number of letters in their names.

_____

_____

2. José bought Christmas presents for his dad, mom, brother, and sister. But before he could give them the presents, his little brother ripped the gift tags off the packages. José cannot remember which package is for which person. What is the probability that José correctly relabels exactly three of the four packages?

_____

_____

3. Six glasses are lined up on the table. The three on the left are filled with water; the three on the right are empty. Make a pattern of full-empty-full-empty-full-empty by moving or touching only one glass.

_____

_____

_____

4. Is it possible for you and a friend to stand behind each other at the same time? Explain your answer.

_____

5. Two mothers and two daughters are walking down the sidewalk when they find three quarters. They pick up the quarters and divide them evenly; each receives one. How is this possible?

_____

6. Christine, Evan, Holly, Johann, and Lindsey are waiting single file in the lunch line (not necessarily in that order). Each has a test in a different subject that afternoon. Their tests are in Bible, English, history, science, and Spanish. Which student has which test this afternoon? (Hint: Figure out the order of the line as you go. Use the chart to eliminate possibilities.)

   a) Johann, who is nearest to the cashier, rides the same bus as the student with the Bible test.
   b) Johann does not have an English test.
   c) Christine is third in line between the student with the science test and the student with the Spanish test.
   d) Lindsey is the sister of the student with the Spanish test. Nobody is standing behind Holly.
   e) Holly is standing behind Lindsey.
   f) The student with the Bible test is at one end of the line.

	Bible	English	History	Science	Spanish
Christine					
Evan					
Holly					
Johann					
Lindsey					

Zoo: Reptiles   **WS 4.3.8B**

Reptile House

Observe the specimens in the reptile house and answer the following questions:

1. What is a herpetologist? _____

_____

2. List all the different colors of reptile scales that you observe as you go through

the reptile house. _____

_____

Crocodilians

The four basic types of crocodilians are alligators, caimans, gharials, and crocodiles.

Alligators

3. Describe the alligator's zoo habitat. _____

_____

4. List three differences between the alligator's zoo habitat and its natural

habitat. _____

_____

_____

5. Describe an alligator's head. _____

_____

6. List two ways that the alligator might use its snout in its natural habitat.

_____

_____

Caiman

7. Describe three characteristics of a caiman.

_____

_____

_____

8. List three similarities between an alligator and a caiman.

_____

_____

_____

Zoo: Reptiles

Gharials

9. Describe a gharial's jaws. _____

_____

10. What does the gharial hunt using this type of jaw? _____

_____

Crocodiles

11. Where do the crocodiles found in this zoo live naturally? Describe their natural habitat and tell which country or countries these animals were brought from.

Habitat: _____

_____

Country or countries: _____

_____

12. List three differences between crocodiles and alligators.

_____

_____

_____

Crocodilian Conservation

13. List two primary reasons why crocodilians are endangered.

_____

_____

Snakes

Venomous Snakes

14. In the chart below, record the common name of any three venomous snakes. Also record the snake's scientific name (two-part Latin name), estimated length, and the part of the world where it is found.

Common Name	Scientific Name	Estimated Length	Natural Habitat

Zoo: Reptiles continued   **WS 4.3.8B**

15. The coral snake and the scarlet king snake look very much alike. Both snakes have black, yellow, and red bands. The coral snake is highly poisonous; the king snake is not. Herpetologists believe the scarlet king snake mimics the poisonous coral snake. Other snakes in the zoo may have these same black, yellow, and red bands. To quickly tell the difference between a poisonous snake with these colors from a nonpoisonous one, remember this rhyme, "Red next to black, venom lack. Red next to yellow, kill a fellow." A snake that has red bands next to yellow bands is poisonous. A snake with black bands next to red bands is not poisonous. In the chart, list all the snakes in the zoo with red, yellow, and black bands, and decide whether or not they are poisonous.

Snake Name	Poisonous?

16. Vipers and pit vipers are heavy bodied, venomous snakes that ambush their prey. They usually lie in wait for their prey, and they often have beautiful camouflage for hiding themselves. The most famous type of pit vipers are the rattlesnakes that live in North and South America. In the chart below, write the common name, scientific name, and camouflage patterns and colors of any three pit vipers.

Common Name	Scientific Name	Scale Patterns and Color

17. What is the shape of a pit viper's head? Why does the head have this shape?

_____

_____

Zoo: Reptiles

Constrictors

18. In the chart, record the following for any three constrictors: common name, scientific name, estimated length, and the part of the world where it is found.

Common Name	Scientific Name	Estimated Length	Natural Habitat

19. How does the size of constrictors compare with that of venom injectors?

_____

_____

_____

20. Why would it be an advantage for a constrictor to be large? _____

_____

_____

21. What is the natural habitat of an anaconda? _____

_____

_____

22. How large can anacondas grow? _____

_____

_____

Lizards

23. Name and describe three lizards in the zoo.

_____

_____

_____

24. What is the largest lizard in the world? _____

25. What is the largest lizard in the zoo? _____

Zoo: Reptiles continued

Turtles

26. What is the main difference between a turtle and a tortoise? _____

_____

27. Record the name, weight, and age of the largest tortoise in the zoo.

Name	Weight	Age

Tuatara

28. If the zoo houses a tuatara, list three of its physical features. _____

_____

Zoo: Birds

Birdhouse

1. Sit quietly for 10 minutes. Write down all of the different kinds of birdsongs

you hear. _____

_____

_____

2. Write down all the different feather colors you see. _____

_____

_____

3. Why do you think many birds have brightly colored feathers? _____

_____

_____

4. Describe the temperature and humidity in the birdhouse. Why do you
think the zookeepers maintain this type of temperature and humidity?

_____

_____

5. Which is your favorite bird at the zoo? Why? _____

_____

_____

Hummingbirds

6. How fast does a hummingbird beat its wings? _____

_____

7. Although hummingbirds occasionally eat small insects or spiders, most
hummingbirds feed primarily on flower nectar. What do the hummingbirds in

the zoo eat? _____

_____

8. The ruby-throated hummingbird is the only hummingbird species that
breeds in the eastern United States. Bird watchers place sugar water
in red feeders to attract this bird to feed. Why do they use red feeders?

_____

9. Watch a hummingbird for 10 seconds. Estimate the distance that it travels in that time. _____

_____

Tropical Birds

10. List four differences between tropical birds and birds found around your house or school. _____

_____

_____

_____

11. Look closely at a parrot's beak. Parrots have a modified beak that combines two types of beaks that you have learned about in class. What are these two types of beaks? _____

_____

12. What does the parrot eat? How does its beak help it to eat this food? _____

_____

13. Why are the populations of many tropical birds endangered? _____

_____

Birds of Prey

14. List three characteristics all birds of prey share.

_____

_____

_____

15. In the chart below, record the common name and scientific name for any three birds of prey kept by the zoo. Also record where the bird lives in the wild.

Common Name	Scientific Name	Natural Habitat

Zoo: Birds continued   **WS 4.3.8C**

16. Describe five physical features of your favorite bird of prey.

_____

_____

_____

_____

_____

Flightless Birds

17. Estimate the height of the tallest ostrich. _____

_____

18. Explain how ostriches can be dangerous to people. _____

_____

19. List any other flightless birds in the zoo and where each bird is from.

_____

_____

_____

_____

Zoo: Mammals

Carnivores

Cats

1. When they think of wild cats, many people picture lions out on the savanna, but 75% of all cat species live in the forest. Identify three characteristics of cats that enable them to survive in forests.

_____

_____

_____

2. The cat family is divided into the big cats and the small cats. There are several species of big cats, including lions, tigers, jaguars, cheetahs, leopards, snow leopards, and clouded leopards. Name the species of big cats kept by the zoo.

_____

_____

3. Several subspecies of tigers live in China, India, North Korea, and Indonesia, but three of these species are now extinct. Only about 3,200 tigers are now living in the wild. Which subspecies of tigers does the zoo keep?

_____

_____

4. In the chart, record the common name, scientific name, and natural habitat of any five small cats.

Common Name	Scientific Name	Natural Habitat

5. Most cats are not hunted by other animals. The only real threat to many cats is humans. List three reasons why people hunt cats.

_____

_____

_____

Zoo: Mammals

Dogs

6. Wolves, coyotes, jackals, foxes, and African wild dogs are all classified as dogs. Which types of dogs are kept at the zoo? _____

_____

Bears

7. The species of bears are brown/grizzly bears, American black bears, Asian black bears, sun bears, sloth bears, speckled bears, and polar bears. Which species of bears are in the zoo? _____

_____

8. Which was your favorite bear? _____

_____

9. Polar bears are the largest land carnivores on Earth. They live in cold, hostile environments. They can survive these harsh conditions by hunting seals. Observe the polar bears. List four characteristics that enable these bears to hunt and survive in cold climates. _____

_____

_____

_____

_____

Primitive Ungulates

10. Ungulates have hooves, but primitive ungulates do not have true hooves. Elephants and aardvarks are primitive ungulates. Observe one of these animals and describe its feet. _____

11. What are the two different species of elephants? How can you tell the difference between these two different species? _____

_____

_____

12. Why are elephants a threatened species? _____

_____

_____

Zoo: Mammals continued   WS 4.3.8D

Odd-Toed Ungulates

13. Odd-toed ungulates have true hooves and an odd number of toes. They include horses, zebras, tapirs, and rhinos. How do these animals benefit from having

hard hooves? _____

_____

_____

14. Horses are not native to North America, but the Native American tribes of the Great Plains used them to hunt buffalo. Wild horses can still be found in the western part of North America. How did horses come to the New World?

_____

_____

15. Discuss four differences between horses and zebras.

_____

_____

_____

_____

16. Describe a tapir. _____

_____

Even-Toed Ungulates

17. Any animal with hooves that is not an odd-toed ungulate is an even-toed ungulate. Even-toed ungulates are found on every continent except Antarctica. Many familiar animals are even-toed ungulates. There are many species of

even-toed ungulates. Name as many as you can. _____

_____

_____

_____

Marsupials

18. Marsupials are mammals with pouches to protect their young and they include the opossum, bandicoot, ringtail, kangaroo, wallaby, koala, and wombat. List

the marsupials kept by the zoo. _____

_____

_____

Bats

19. Why are bats important? _____

_____

20. What percentage of all mammal species are bats? _____

_____

21. Describe a bat's wings. _____

_____

22. Vampire bats are a misunderstood mammal. Although they do hunt on moonless nights in search of blood, they feed on cattle. Rarely do they attack humans. They do not suck blood from cattle. Instead, they lap it up with their tongues after cutting open a small wound. One feeding will last between 8 and 18 minutes, during which time a vampire bat may ingest up to 40% of its own weight in blood. This is a small blood loss for a cow. The main danger from vampire bats is that they often spread rabies. Observe the vampire bats for several minutes.

Write four characteristics of vampire bats.

_____

_____

_____

_____

Zoo: Amphibians

Amphibians

1. List all the different colors of frogs, toads, and salamanders that you see.

_____

_____

_____

2. What are three differences between frogs and toads?

_____

_____

_____

3. In the chart, list three species of amphibians capable of secreting poison of some kind. Write the common name, scientific name, and a fact about its poison.

Common Name	Scientific Name	Poison Fact

4. What are four reasons that amphibians are endangered around the world?

_____

_____

_____

_____

Zoo Trip Summary

1. List three reasons why zoos are important. _____

_____

_____

_____

2. Name two primary reasons that many species of animals are endangered around the world.

_____

_____

3. List five endangered vertebrates kept by the zoo. Include the natural region where each animal lives and why it is endangered.

Endangered Vertebrate	Natural Habitat	Reasons for Being Endangered

4. Modern zoos are replacing cages with iron bars with larger compounds that have a more natural setting. Compare the behavior of animals that you saw in cages with animals you saw in newer compounds. _____

_____

_____

5. Zookeepers often hide the animals' food to force them to search for it instead of just putting their food in pans. Why do you think they do this? _____

_____

_____

6. List several advantages of these new compounds and hiding the animals' food.

_____

_____

Zoo Trip Summary

7. Compare the behavioral differences of the mammals and birds with the reptiles and amphibians. _____

_____

_____

Aquarium

Jawless Fish

1. The three basic types of jawless fish are lampreys, slime eels, and hagfish. List the different species of jawless fish kept in the aquarium. _____

_____

_____

2. What are three characteristics that all jawless fish share?

_____

_____

_____

Sharks, Rays, and Skates

3. List three general characteristics of rays.

_____

_____

_____

4. What is the largest type of ray? _____

5. Name the different species of rays kept in the aquarium. Record their common name and scientific name.

Common Name	Scientific Name

6. Wait for a ray to swim up against the glass and describe what its underside looks like. _____

_____

_____

7. What species of sharks are kept at the aquarium? Record their common name and scientific name.

Common Name	Scientific Name

8. As a shark passes by, closely examine its skin, and describe what it looks like.

_____

_____

_____

9. As a shark passes by you, closely examine its teeth. Describe them. _____

_____

_____

10. How are a shark's jaws and teeth uniquely created for hunting? _____

_____

_____

Bony Fish
Coral Reef Fish

11. Name the different colors of fish at the coral reef display. _____

_____

_____

_____

12. In what geographical region of the world do coral reefs thrive? _____

_____

13. Coral reefs are often called *underwater rain forests*. How did they get this

name? _____

_____

_____

Aquarium continued   **WS 4.3.8G**

14. How do fish use color to defend themselves? _____

_____

_____

_____

_____

15. Name two predators living on the coral reef. _____

_____

16. Moray eels are constantly opening and closing their mouths, but they are not

doing this as a warning. Why do you think they do this? _____

_____

_____

_____

17. What do the nonpredatory fish on the reef eat? _____

_____

18. Look closely at a butterfly fish. Many butterfly fish have large black dots on
their tails, but their eyes are hidden by a black strip. Explain the advantage of
this color pattern. (Hint: Butterfly fish have sharp spines on their dorsal fins

that point toward their tail.) _____

_____

_____

_____

19. What type of coral reef fish makes sand for Caribbean beaches? How does it

do this? _____

_____

20. List several differences between freshwater fish and saltwater fish. _____

_____

21. Why do marine biologists intentionally sink old ships in tropical waters?

_____

_____

Aquarium

Mangroves

22. Explain several reasons why mangroves are important to fish. _____

_____

Bottom-Dwelling Fish

23. List several differences between flounder and most other bony fish. _____

_____

24. Most bottom-dwelling fish do not have a swim bladder. Why not? _____

_____

Schools

25. List several species of fish that school together. _____

_____

26. Why do you think fish school together? _____

_____

27. Why do you think most predators like barracudas and tarpons do not hunt schooling fish until dawn or dusk? (Hint: Think about the sun's rays at those

times of the day.) _____

_____

Marine Reptiles

28. Sea turtles are the most common marine reptiles. Explain how sea turtles are

different from tortoises. _____

_____

29. What are three reasons why many sea turtle species are endangered?

_____

_____

_____

Sea Mammals

30. List the different types of sea mammals you see at the aquarium. _____

_____

_____

_____

Aquarium continued   WS 4.3.8G

31. How are seals uniquely suited for living in cold water? _____

_____

_____

32. Most seals can stay underwater for 20–30 minutes. Weddell seals can stay underwater for over an hour. Time how long you can hold your breath and compare it to the seals' diving time. _____

_____

33. What is the difference between a dolphin and a porpoise? _____

_____

34. What is a cetacean? _____

_____

35. What types of cetaceans do you see at the aquarium? _____

_____

36. List four facts about a sea mammal that you see at the aquarium.

_____

_____

_____

_____

Name: _____ Date: _____

Lab 5.1.1A Human Cell Shape

QUESTION: How are human cells shaped?

HYPOTHESIS: _____

_____

EXPERIMENT:

You will need:	• strand of human hair
• prepared microscope slides of different tissues and organs	• microscope

Steps:

1. Choose a microscope slide of human body cells. Place a strand of your hair on the slide next to the coverslip of the prepared slide for a size reference of what you see at each magnification.

2. Set up the microscope to view the cells on your slide with the low-power objective. Slowly move the slide to find the best view of the cells. Label the type of cells being observed. Sketch your observations.

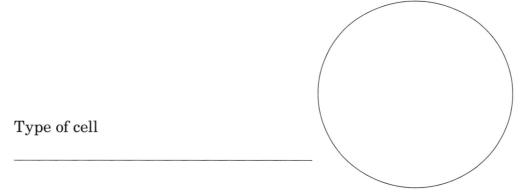

Type of cell

_____

3. Carefully move the hair next to the cells. Sketch a picture to compare the size of the hair with the cells on the slide.

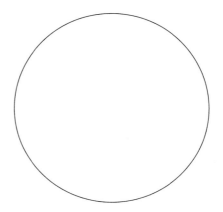

Lab 5.1.1A Human Cell Shape

4. View the cells at higher magnification. Sketch your observations, labeling the cell parts you can identify. Record the magnification at which you viewed the slide.

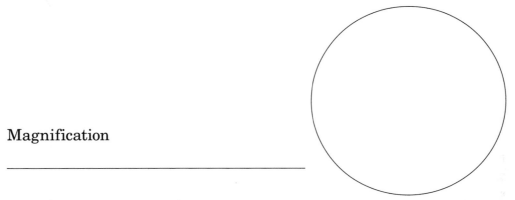

Magnification

_____

5. Repeat the process using the low-power objective with slides from different tissues or organs. Include your hair as a comparison.

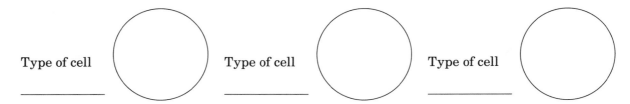

Type of cell

_____

Type of cell

_____

Type of cell

_____

6. Observe the same cells under higher magnification. Sketch your observations and label cell parts. Record the magnification used.

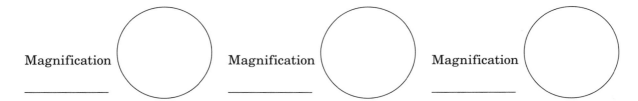

Magnification

_____

Magnification

_____

Magnification

_____

ANALYZE AND CONCLUDE:

1. Why do the cells have different shapes? _____

_____

2. Compare the similarities and differences of the cells of the different organs or

tissues that you observed. _____

_____

_____

3. Estimate how many cells would line up across the diameter of the hair. _____

Lab 5.1.2A Exercise and Pulse Rate

QUESTION: How does exercise affect your pulse rate?

HYPOTHESIS: _____

_____

EXPERIMENT:

You will need:	• stopwatch

Steps:

1. Find your carotid pulse. Have your partner use a stopwatch to time 15 seconds while you count the number of beats. You may feel two beats occurring closely together. These two beats count as one. Multiply the number of beats by 4 and record the number in the Resting portion of the Pulse Rate Data Chart below. This is your pulse rate for one minute.
2. Run in place for 1 minute while your partner times you. Stop immediately and take your pulse for 15 seconds. Multiply that number by 4. Record your pulse on the chart.
3. Sit down right away and have your partner time you as you rest for one minute. After this rest period, take your pulse immediately following the same procedure as before. Record your pulse rate for 1 minute on the chart.
4. Rest for 3 more minutes while your partner times you. Take your pulse again and record it on the chart.
5. Switch places with your partner and repeat Steps 1–4.

Pulse Rate Data Chart

Activity	
Resting	
After Excercise	
1 Minute Recovery	
3+ Minutes Recovery	

ANALYZE AND CONCLUDE:

1. What happened to your pulse rate after exercise? Explain why. _____

_____

_____

2. What happened to your pulse rate after resting for 1 minute and then 3 minutes?

Explain why. _____

_____

_____

Lab 5.1.2A Exercise and Pulse Rate

3. Use the data you recorded on the Pulse Rate Data Chart to create a bar graph. Give the graph a title, and label the *x*- and *y*-axes. Mark numeric increments.

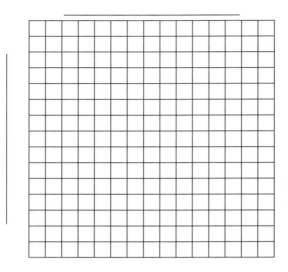

4. Compare your graph results to those of three other classmates. _____

_____

5. Compare the results with your hypothesis. _____

_____

6. How do your muscle cells get oxygen? _____

_____

_____

7. Why does your pulse rate increase when you exercise? _____

_____

_____

8. Athletes who train often have very short recovery periods. How do you think you could shorten your recovery period? _____

_____

_____

9. What conclusion can you draw about the relationship between pulse rate and physical activity? _____

_____

Name: _____ Date: _____

Lab 5.1.2B Sheep Heart Dissection

QUESTION: How is a sheep heart different from a human heart?

HYPOTHESIS: _____

_____

EXPERIMENT:

You will need:	• dissection pan	• metric ruler	• probe
• sheep heart	• scalpel	• gloves, 2 per student	

Steps:

The sheep heart is very similar to the human heart in both size and structure. Study the illustrations of the ventral and dorsal views on **BLM 5.1.2B Sheep Heart External Anatomy**. Use Figures A–H on **BLM 5.1.2C Dissection Stages** to guide you through the dissection process. Be precise with your incisions. Otherwise, you may not obtain the desired results.

1. Put on gloves and place the sheep heart in the dissection pan. Identify the

   ventral and dorsal sides. Describe the differences. _____

   _____

2. Locate the following blood vessels and areas of the heart. Check off each one as you find it. Use Figure A to ensure the proper position of the vessels. Have your teacher check this before continuing.

   _____ aorta _____ apex _____ coronary artery _____ superior vena cava

   _____ brachiocephalic artery _____ pulmonary vein _____ pulmonary artery

   _____ right/left atria _____ right/left ventricles

3. Position the heart with the ventral side showing as seen in Figure B.

4. Place the scalpel about 2 cm to the left of the coronary artery and at the base of the pulmonary artery. Measure with a ruler as in Figure C. Carefully make a shallow incision running parallel to the coronary artery down as far as you can go without turning the heart over. Remember to stay parallel through the entire incision. Gently open the heart as seen in Figure D. Look inside. The pulmonary valve and the tricuspid valve should be visible. Describe what you see in detail.

   _____

   _____

5. Look at Figure E for details. Keeping parallel to the coronary artery, cut around to the dorsal side. Notice the coronary artery curves upward and runs vertically. Stop the incision just to the right of the dorsal coronary artery. Find the superior vena cava (SVC). Staying parallel to the coronary artery, cut downward through the SVC until this incision meets the first incision.

Lab 5.1.2B Sheep Heart Dissection

6. Very gently spread the sides apart, as shown in Figure F. The whole right interior should be visible now. Describe the difference between the walls of the right atrium (RA) and the right ventricle (RV). _____

_____

Explain why there is a difference. _____

7. Find the tricuspid valve located between the RA and the RV. What is the purpose of this valve? _____

8. Place the heart back on the pan with the ventral side up. Find the pulmonary vein (PV), which is located on the top toward the back of the left atrium, above and to the left of the coronary artery. Make an incision through the PV straight down to the apex. Use Figure G as a guide.

9. Gently open the left side of the heart, as shown in Figure H and observe the walls of the left ventricle (LV). Describe the difference between the walls of the LV and the RV. _____

Why is there a difference? _____

10. Look for the bicuspid valve between the LA and the LV. Describe what you see.

_____

11. Find the aorta. Keep the left side of the heart open. Gently slide a probe into the aorta until you see the tip. Into what chamber does it empty? _____

12. Slide the probe down the pulmonary vein. Where does it go? _____

13. Slide the probe down the superior vena cava. To what chamber does it go? _____

_____

14. Compare your observations in Exercises 11–13 with what you have previously learned about the structure of the heart. Do your observations support this information? _____

15. What was the most interesting thing you found about this dissection? _____

_____

16. Observe the interior structure of the sheep heart again. Describe how the heart is perfectly suited for its purpose. Reflect on how this deepens your faith in God the Creator. _____

_____

Help a Heart

Research to come up with a recovery plan for a heart attack patient.

1. Describe the patient.

Age _____

Gender _____

Lifestyle (Smoker? Exercise?) _____

Family history of heart disease _____

Blood pressure (High? Normal?) _____

2. What would be the recommendations of

a dietician (making diet recommendations) _____

_____

_____

_____

a personal trainer (making exercise routine and fitness goals) _____

_____

_____

_____

_____

a counselor (making recommendations for how to deal with stresses in life) _____

_____

_____

_____

_____

Heart Diagram   **WS 5.1.2B**

Label the following parts of the heart:

Right and left atrium, right and left ventricle, aorta, superior vena cava, inferior vena cava, pulmonary veins, pulmonary artery, septum, and two valves.

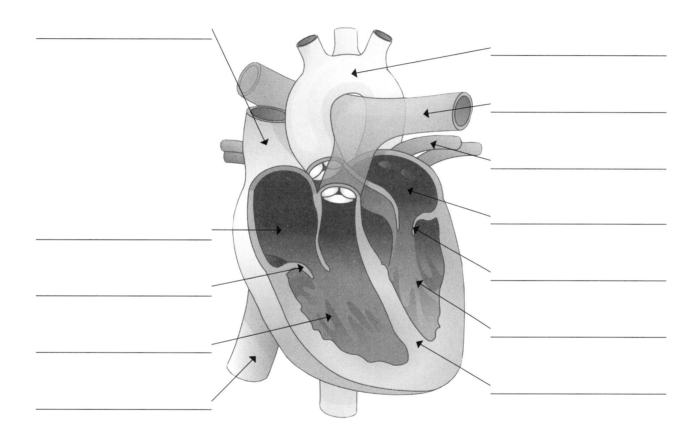

Digestive System Diagram

Label the mouth, esophagus, stomach, liver, gallbladder, pancreas, small intestine, and large intestine.

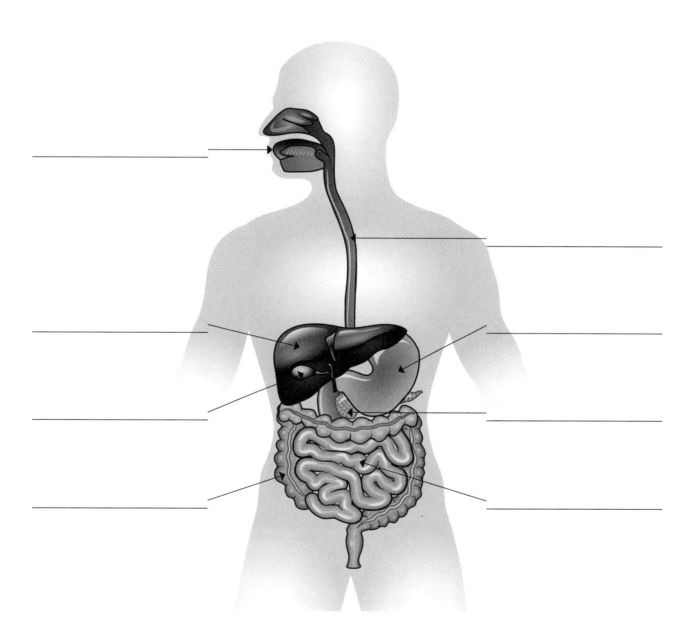

Lab 5.2.1A Daily Breath

QUESTION: How much air do I breathe each day?

HYPOTHESIS: _____

_____

EXPERIMENT:

You will need:	• plastic milk or juice jug	• piece of plastic tubing
• large capacity graduated cylinder	with a cap	
• large basin	• water	

Steps:
1. Use the graduated cylinder to measure as you fill the plastic jug completely with

   water. Record the volume of water in the jug. _____ Secure the cap.
2. Fill the basin with 6–8 cm of water.
3. Turn the jug upside down over the basin so the neck of the jug is submerged in the water.
4. Without lifting the neck of the jug out of the water, remove the cap of the jug. Insert one end of the plastic tubing into the jug, keeping the neck of the jug in the water.
5. Hold the other end of the tubing in your hand. Breathe in normally and exhale once into the end of the tubing that you are holding. The air in your lungs will displace some of the water from the jug. The displaced water will go into the basin.
6. Remove the tubing and replace the cap on the jug, being careful not to lift the neck of the jug out of the water.

Lab 5.2.1A Daily Breath

7. Turn the jug right side up. Remove the cap and measure the remaining water, using the graduated cylinder. Subtract this measurement from the original

   volume of water to calculate the volume of air that you exhaled. _____

8. Breathe in a relaxed fashion and count how many breaths you take in 1 minute.

   Record the number of breaths you took in 1 minute. _____

ANALYZE AND CONCLUDE:

1. Calculate how much air (volume) you breathe in a day of normal breathing.

Volume of air in one breath _____

Number of breaths per minute _____

Volume of air breathed per minute _____

Number of minutes per day (24 hrs) _____

Total volume of air breathed in one day _____

2. Compare your hypothesis to the total volume of air you breathe in one day. _____

_____

Research to complete the following exercises:

3. What percentage of air is oxygen? _____

4. Based on the total volume of air you breathe in one day, calculate the amount of

   oxygen you take in. _____

Urinary System WS 5.2.2A

1. Label the following: kidney, aorta, vena cava, bladder, ureter, urethra, renal artery, renal vein, renal corpuscle, glomerulus, and loop of Henle.

Kidneys and Bladder   **Nephron**

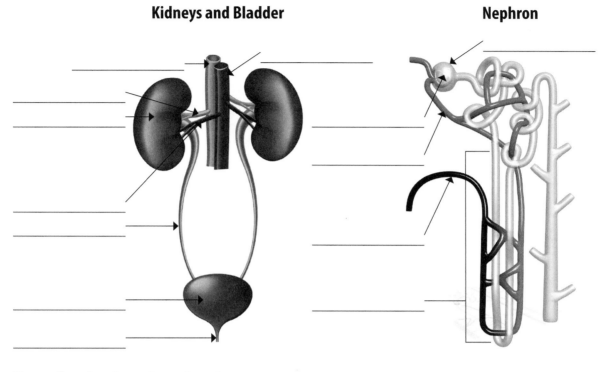

2. Describe the function of each.

a) kidney _____

_____

b) bladder _____

_____

c) ureter _____

_____

d) urethra _____

_____

e) nephron _____

_____

f) glomerulus _____

_____

Tracking UVs

1. Chart the UV index as given by the weather service in your area for one week.

UV Index Chart

Days	Date	UV Index
Day 1		
Day 2		
Day 3		
Day 4		
Day 5		
Day 6		
Day 7		

2. Create a line graph from the data in the UV Index Chart. Title the graph and label the *x*-axis and the *y*-axis.

3. Look up the UV index scale. The scale ranges from 0 to 11+. List the ranges for low, moderate, high, very high, and extreme. _____

_____

4. List any days that were in the *high, very high, or extreme* range. _____

_____

5. What are some precautions you can take on days with high UV index readings?

_____

_____

Lab 5.3.1A Bones and Acid

QUESTION: What will happen to a bone placed in hydrochloric acid?

HYPOTHESIS: _____

_____

EXPERIMENT:

You will need:	• 2 long chicken or turkey bones	• distilled water
• masking tape	(boiled, all meat removed)	• forceps
• marker	• 1N hydrochloric acid (enough to cover	• paper towel
• 2 glass jars with lids	the bone)	

Steps:
1. Use masking tape and a marker to label the jars *Jar 1* and *Jar 2*.
2. Observe the color and flexibility of the two long bones. Record your observations on the chart below.
3. Place one bone in *Jar 1* and fill the jar with distilled water. Place the other bone in *Jar 2* and fill it with enough 1N hydrochloric acid to cover the bone. Cover both jars with the lids.
4. Allow the jars to sit undisturbed 12–24 hours. Carefully remove the bones with forceps. Gently rinse the bones in water a few times and then dry the bones with the paper towel.
5. Observe and record what happened to the color and flexibility of the bones.

		Before	**After**
Jar 1 Distilled Water	**Color:**		
	Flexibility:		
Jar 2 Hydrochloric Acid	**Color:**		
	Flexibility:		

ANALYZE AND CONCLUDE:

1. What are bones made of? _____

_____

2. What did the hydrochloric acid do to the bone? _____

_____

3. Why are the minerals in bones important? _____

_____

Lab 5.3.2A Chicken Legs

QUESTION: How are chicken legs structured?

HYPOTHESIS: _____

_____

EXPERIMENT:

You will need:	• hand lens	• scalpel
• latex or nonlatex gloves	• dissection tray	• safety goggles
• raw chicken leg quarter with skin	• scissors	

Steps:

1. Put on gloves. Observe the uncooked chicken leg quarter (drumstick and thigh attached). Look closely at the skin. Can you see where the feathers were once attached?

   _____

2. Remove the skin and observe the bundles of muscle fibers that make up the various muscles. Observe the muscle bundles with the hand lens. Move the joint and observe the movement of the various muscle groups. Sketch the leg, labeling muscle bundles, tendons, fat, and bone.

3. Place the leg in the dissecting pan. Dissect the leg by separating the muscle groups with scissors, following each muscle group to the point of its insertion into the bone. Observe how the tendon attaches to both muscle and bone. Use the scalpel to try and separate an individual muscle fiber.

4. One by one, disconnect the muscles from the bone, leaving only the two bones connected at the joint. Locate the ligaments. Sketch the joint and label the ligaments that hold the joint together.

5. Carefully separate the bones and inspect the interior of the joint. Carefully break the largest bone. Do not crush it. Observe the red jellylike tissue inside the bone.

ANALYZE AND CONCLUDE:

1. Describe how a chicken leg is structured. _____

2. What is chicken meat? _____

3. What is the red jellylike tissue inside the bone? _____

Skeletal System

1. Label the *cranium*, *sternum*, *scapula*, *humerus*, *ulna*, *radius*, *femur*, *tibia*, *fibula*, and *vertebrae*. Draw brackets to indicate the axial skeleton and appendicular skeleton.
2. Label the joints: *ball and socket*, *hinge*, *pivot*, and *gliding*.

Human Skeleton　　　　　　　　　　　　　**Human Joints**

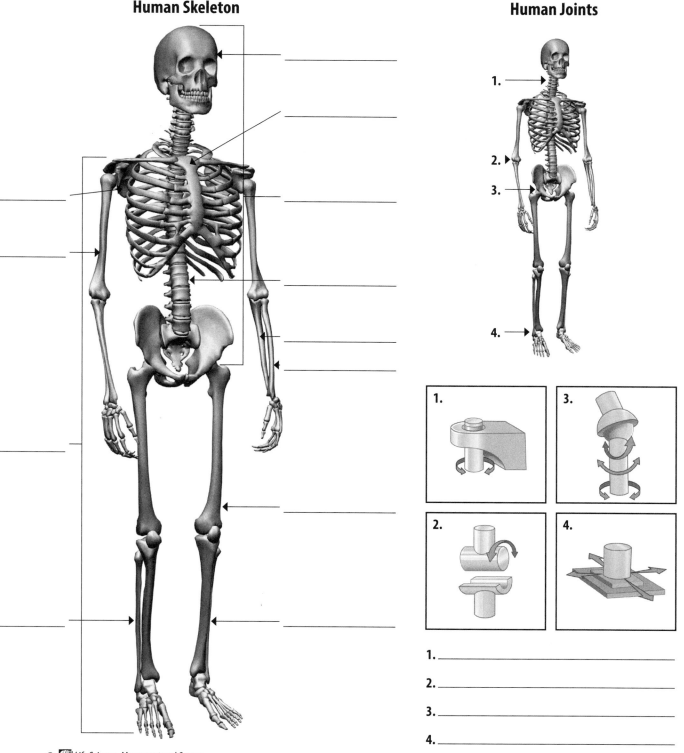

1. _____

2. _____

3. _____

4. _____

Brainy Activities

Label the following parts of the brain: *cerebrum, cerebellum, brain stem, thalamus,* and *hypothalamus.*

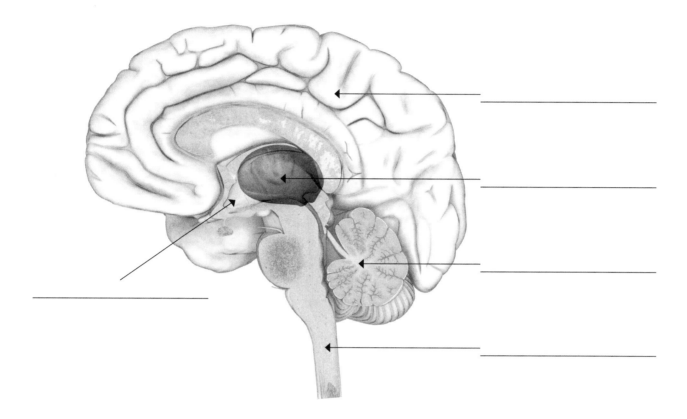

Make a list of 10 things that you do throughout the day (involuntary actions and voluntary actions). Indicate which part of the brain controls that function.

1. _____

2. _____

3. _____

4. _____

5. _____

6. _____

7. _____

8. _____

9. _____

10. _____

Eyes and Ears

1. Label the *iris, cornea, retina, lens, pupil,* and *optic nerve.*

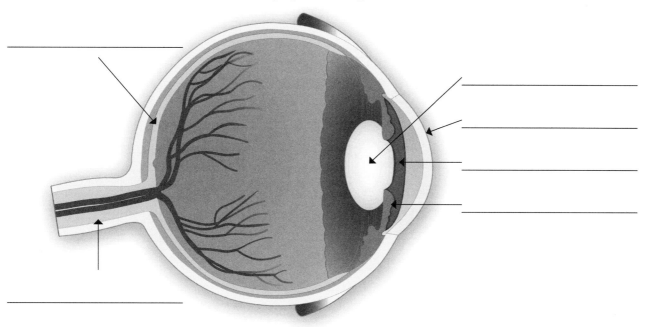

2. Label the *outer ear, middle ear, inner ear, tympanic membrane, ear canal, anvil, hammer, stirrup, eustachian tube, cochlea,* and *semicircular canals.*

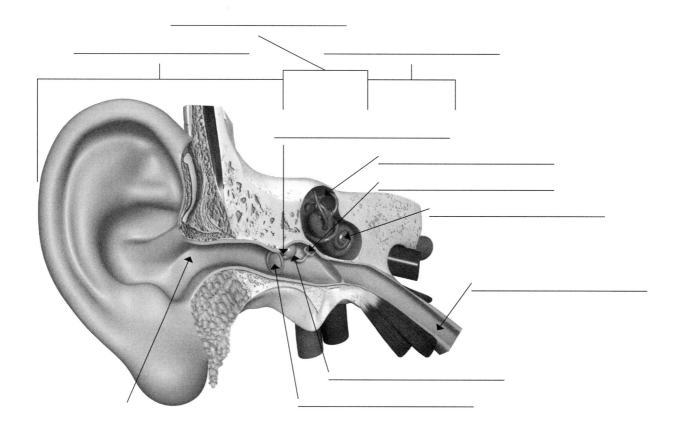

Endocrine Glands   **WS 5.4.1A**

Draw and label the endocrine glands.

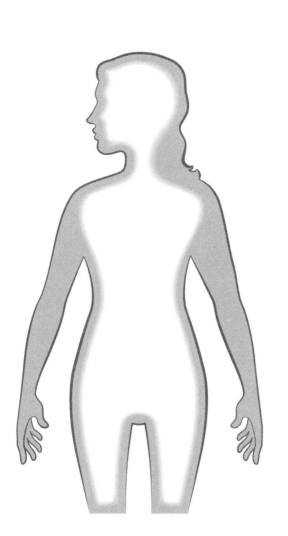
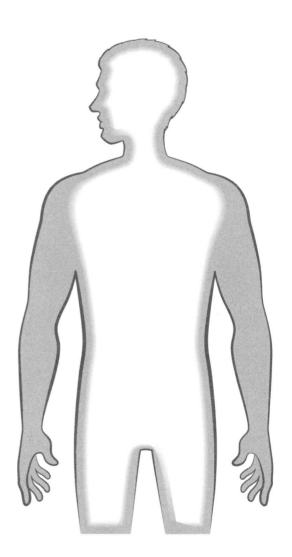

Circadian Rhythms

Keep a bean plant, a shamrock plant, and a prayer plant in an environment with natural light and a constant temperature. Water the plants regularly. Observe the position of the leaves of each plant at regular times each morning, afternoon, and evening for three days. For example, observe them at 9:00 AM, 3:00 PM, and 9:00 PM. Sketch the leaf positions on the data charts.

Bean Plant

Time		
Day 1			
Day 2			
Day 3			

Shamrock Plant

Time		
Day 1			
Day 2			
Day 3			

Circadian Rhythms

Prayer Plant

Time			
Day 1			
Day 2			
Day 3			

I've Got Rhythm   **WS 5.4.3B**

Your body temperature is a circadian rhythm that you can monitor. Record your temperature every day for three days at 7:00 AM, 10:00 AM, 1:00 PM, 4:00 PM, 7:00 PM, and 10:00 PM. (If you go to bed before 10:00, take your temperature right before you go to bed.) Do not eat, drink, or brush your teeth for 15 minutes before taking your temperature. Read the thermometer carefully to note differences in your temperature.

	7:00 AM	**10:00 AM**	**1:00 PM**	**4:00 PM**	**7:00 PM**	**10:00 PM**
Day 1						
Day 2						
Day 3						

1. What pattern do you observe? _____

_____

_____

2. Graph the results. Then find the average temperatures for each time of day for the class and graph the results.

My Temperatures   **Class Average Temperatures**

_____   _____

My Meal   WS 5.4.4A

Record the foods and the beverages that you had for breakfast or lunch. Use a nutrient database to analyze the nutrients of each food.

Food	Protein	Fat (g)	Fat (% of calories)	Carbohy- drate	Vitamin A	Vitamin C	Thiamine	Riboflavin	Niacin	Iron	Calcium

Balancing Act

Using data on the Internet, design a balanced diet for one day. It should include 2,000 calories (no more than 30% fat) and recommended servings of the five food groups. Write down any added fats and your planned water intake as well.

Servings	Fruits	Vegetables	Grains	Proteins	Dairy	Fats	Water
	2	2–3	6	6	3	0–1	8
Breakfast							
Snack							
Lunch							
Snack							
Dinner							

Drug Study

You will be assigned one of the drug groups (cannabis, stimulants, depressants, hallucinogens, or narcotics) to research and present to the class. Use the questions below to guide your research.

Drug Group: _____

1. Which drugs fall into this group? (Include the proper names and nicknames.)

_____
_____
_____
_____
_____

2. In what forms are the drugs found? How are they used (smoked, injected,

swallowed, or inhaled)? _____

_____
_____
_____
_____
_____
_____

3. What are the short-term effects of the drug on the body? _____

_____
_____
_____
_____

4. What might the long-term effects be? _____

_____
_____
_____
_____

Name: _____ Date: _____

Lab 6.1.2A Fingerprinting

QUESTION: What type of fingerprints do I have—whorls, arches, or loops?

HYPOTHESIS: _____

_____

EXPERIMENT:

You will need:	• several pieces of white paper	• hand lens
• soft lead pencil (#2) or charcoal pencil	• clear removable tape	• metric ruler

Steps:
1. Wash and dry your hands.
2. On a piece of paper, write a heading for each finger and the thumb of your left hand. Do the same on another piece of paper for your right hand. Leave enough room below each heading for a fingerprint.
3. Using the pencil, shade in a 3 cm square on another piece of paper.
4. Rub the pad of one of your fingers over the shaded square. Carefully place a piece of tape onto the blackened finger so that the tape comes in contact with the entire fingerprint. Carefully peel the tape away. Stick the tape under the heading for the proper finger. Repeat the process for each finger, adding a bit of fresh graphite from the pencil to the paper before taking each fingerprint.
5. Wash and dry your hands.
6. Use a hand lens to carefully inspect each fingerprint. Decide which of the three basic fingerprint types the fingerprint looks most like.

Whorl   **Arch**   **Loop**

7. Record what type of fingerprint is displayed by each finger.
8. Add up the number of fingers that display loops, whorls, and arches and record the number on the Fingerprint Data Chart. Compare your fingerprint types with those of your partner or group.
9. Add your data to the class data chart. Record the totals for the class on your Fingerprint Data Chart.
10. Calculate the percentage of fingerprints in the class that fall into each category.

Lab 6.1.2A Fingerprinting

Fingerprint Data Chart

	Whorls	Arches	Loops
Mine			
Class Total			
Class %			

ANALYZE AND CONCLUDE:

1. What did you learn about your own fingerprints? _____

_____

2. How did your type of fingerprints compare with the class totals? _____

_____

3. Which was the most common type of fingerprint in your class—whorl, arch, or loop?

_____

_____

The least common? _____

_____

4. Research to find the percentages of fingerprint types in the general population.

_____

_____

5. How does your class percentage match with the general population? _____

_____

Lab 6.1.4A Different Ways to Mutate

QUESTION: What are some different ways in which a mutation can occur?

HYPOTHESIS: _____

_____

EXPERIMENT:

| **You will need:** | • paper DNA models previously | • scissors |
| • tape | constructed | |

Steps:

1. Using the DNA model, put together a single strand that has the following sequence: AGT TAC CAT GGC TCT.

2. Complete the double helix by building the complementary strand of DNA, which is the strand that contains the gene to be read. It is called *the sense strand*.

   Record that gene sequence. _____

3. Unzip the helix and read the DNA sense strand (from left to right as the phosphate-sugar backbone is along the bottom). Convert it into mRNA using the remaining nucleotide pieces. Record the RNA sequence. _____

   _____

4. Finally, convert the mRNA into a protein (see the Genetic Code Triplets in the Student Edition) and record the sequence. _____

5. Explore the effects of a "deletion" mutation in which a nucleotide is missing. Remove the fourth nucleotide (the nitrogen base, phosphate, and sugar) on the original strand. Record the new sequence. _____

   _____

   Adjust the remaining strands—the DNA sense strand and the mRNA—by removing the fourth nucleotides. Record the sequences for the sense strand

   _____

   and the mRNA. _____

6. Convert the mRNA into amino acids. Record the sequences. _____

   _____

Lab 6.1.4A Different Ways to Mutate

7. What changes were made? _____

_____

8. Use this new DNA sequence that forms a gene to complete the following activities:

ATG GGT CGT ACG ACC GGT AGT TAC TGG TTC AGT TAA

a) Convert the gene into mRNA. Record the sequence. _____

_____

b) Then, convert that gene into the amino acid sequence of the protein coded by

this gene. _____

c) Investigate what effect a "substitution" mutation in the DNA would have. Count from the left to the fifteenth nitrogen base; then substitute a T for the C. First, convert the DNA into RNA.

d) Then, convert the new strand into protein. Are there any changes in the

protein? If so, what are they? _____

_____

e) Investigate the effects of an "insertion" mutation by inserting a two-nucleotide sequence of AT between the sixth and seventh DNA nucleotides. First, convert

the DNA into RNA. Record the RNA sequence. _____

_____

Now change the new strand into a protein. Are there any changes in the

protein? If so, what are they? _____

_____

9. What are some of the ways that a mutation can occur? _____

_____

Class Traits

Your teacher will display some images. Observe the different traits. Dominant traits are listed in the left-hand column. Survey your classmates for the following traits and write in the number of students displaying each one. After collecting your data, convert your data to percentages. For example, if 15 out of 30 students show brown eyes, record that information as 50%.

Thumbs

Total students _____   Total students _____

Number with straight thumbs _____   Number with bent thumbs _____

Percentage with straight thumbs _____   Percentage with bent thumbs _____

Hair

Total students _____   Total students _____

Number with curly hair _____   Number with straight hair _____

Percentage with curly hair _____   Percentage with straight hair _____

Eye Color

Total students _____   Total students _____

Number with brown eyes _____   Number with nonbrown eyes _____

Percentage with brown eyes _____   Percentage with nonbrown eyes _____

Summarize the results above using complete sentences. Were there more dominant or recessive traits in your class? _____

_____

_____

_____

_____

_____

_____

_____

_____

_____

_____

Traits of an Imaginary Creature   WS 6.1.3A

1. Design an imaginary creature. Choose five traits and determine a dominant and a recessive phenotype for each trait. (For example, pointy ears—dominant, round ears—recessive; purple nose—dominant, green nose—recessive.) List your traits

   below. _____

   _____

   _____

   _____

2. Create a genotype for each phenotype. _____

   _____

   _____

   _____

3. Choose one phenotype and decide on the genotypes of a male and a female.

   _____

   _____

   _____

   _____

4. Draw a Punnett square using the genotypes in Exercise 3 to determine the probability of the phenotype in the offspring.

5. What proportion of the offspring would likely have the dominant trait? _____

6. What proportion would likely express the recessive trait? _____

7. What proportion would be homozygous? _____

8. What proportion would be heterozygous? _____

Name: ______________________ Date: __________

Bradykinin

WS 6.1.4A

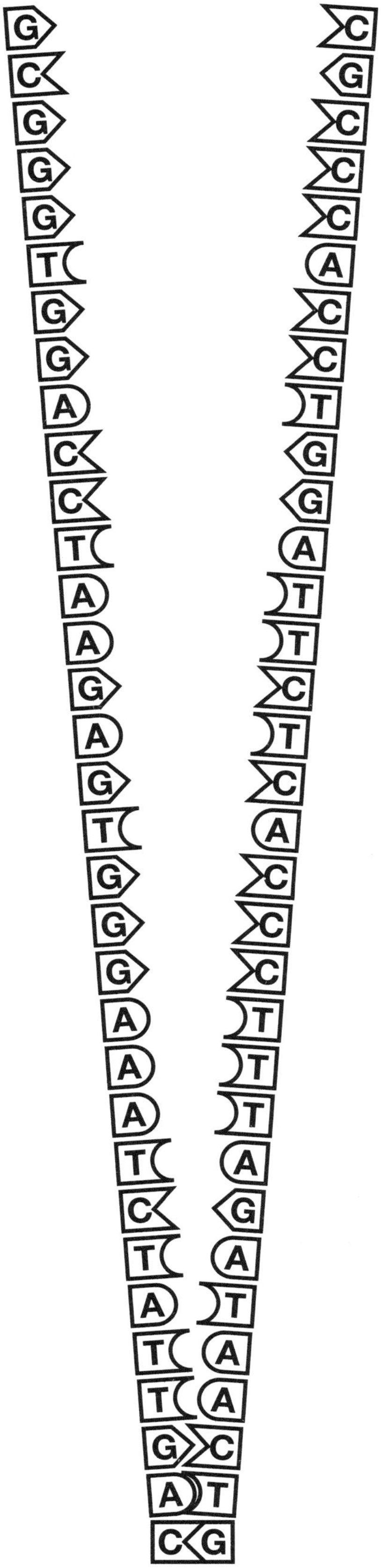
G C G G G T G G A C C T A A G A G T G G G A A A T C T A T T G A C
C G C C C A C C T G G A T T C T C A C C C T T T A G A T A A C T G

Convert the information on the gene into the protein bradykinin. Bradykinin is a protein in the blood that helps regulate blood pressure.

What is the sequence of amino acids that make up bradykinin? (Hint: You must first determine what RNA molecule would be made from the DNA molecule.)

RNA molecule:

____

____

____

____

RNA triplet (codon):

____

____

____

____

Amino acid sequence:

____

____

____

____

DNA and RNA WS 6.1.4B

Answer the questions to complete the chart.

	DNA	RNA
What do the letters stand for?		
What is it composed of?		
Where does it work?		
What does it do?		

Protein Synthesis

Complete the flowchart to show the steps of protein synthesis.

Venn Diagram: Mitosis and Meiosis   **WS 6.1.5A**

Mitosis

Shared

Meiosis

Stages and Phases of Mitosis   WS 6.1.5B

Label the stages and phases. Describe the events that occur during each one.

_____

_____

_____

_____

_____

_____

Chromosome Matching

Geneticists match chromosome pairs to create a karyotype. Use what you have learned about karyotypes to pair up the chromosomes pictured.

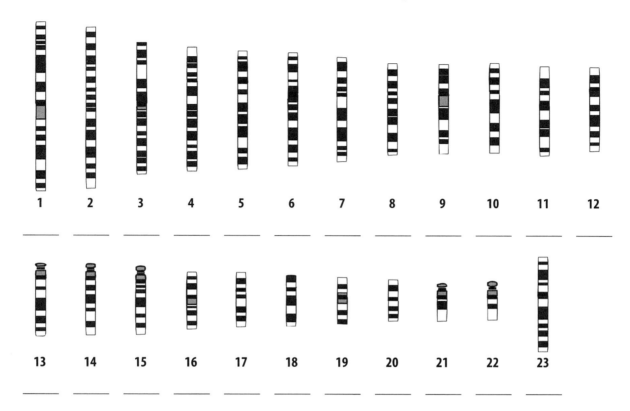

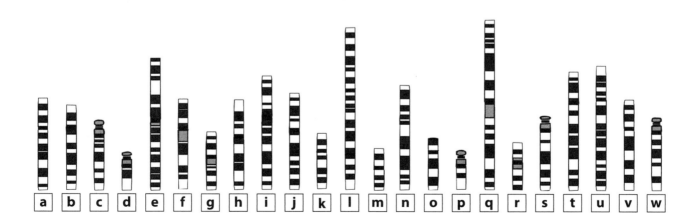

Is the person male or female? _____

How do you know? _____

Genetics

Give an example of the following:

1. trait _____

2. variation _____

3. gamete _____

4. gene _____

5. allele _____

6. dominant trait _____

7. recessive trait _____

8. homozygous _____

9. heterozygous _____

10. genotype _____

11. phenotype _____

Fill in the blanks with the appropriate word.

12. The study of how certain features are passed from parent to offspring is

_____.

13. The passing of traits from parents to their offspring is _____.

14. The fertilization of the egg of one maple tree with the pollen of another maple tree

is an example of _____.

15. The offspring of two different maple trees is a _____.

16. The chemical instructions that translate genetic information make up the

_____.

17. Something that causes mutations to occur is a _____.

18. New gametes are formed through the process of _____.

19. New somatic cells are formed through the process of _____.

20. Color blindness, whose genes are carried on the X chromosome, is a

_____.

21. The total genetic material of an organism is its _____.

Codominance

Read each example below. Write *codominance* or *incomplete dominance* in the blank lines below to indicate what each example represents.

1. Cows can be black, white, or white with black spots. _____

2. Flowers can be red, white, or pink. _____

3. A chicken can have brown feathers, white feathers, or some of each. _____

_____

People have different blood types based on the presence or absence of certain proteins. The alleles that provide the coding for blood type are I^A, I^B, and i for blood type O. Blood types A and B are codominant, and O is recessive. The O blood type indicates the absence of the A and B alleles. Each parent donates one blood type allele to each offspring.

4. What genotypes are possible for the following blood types?
Type A: _____ or _____   Type AB: _____
Type B: _____ or _____   Type O: _____

5. Complete the cross to show that the father is homozygous for type B blood and the mother is codominant type AB.

6. What are the possible genotypes for the children of these

parents? _____

7. What blood type(s) could their children express? _____

_____

8. Is there any blood type that could not be expressed by their

children? If so, what is it? _____

_____

9. Complete the cross to show that the mother and father are both heterozygous for type A blood.

10. What are the possible genotypes for the children of these

parents? _____

_____

11. What blood type(s) could their children express?

_____

12. Is there any blood type that could not be expressed by their

children? If so, what is it? _____

Lab 6.2.1A Manufacturing Insulin

QUESTION: How is genetic engineering used to manufacture human insulin from bacteria?

HYPOTHESIS: _____

_____

EXPERIMENT:

You will need:	• 2 strands blue, green, and red yarn	• 5 labels	• scissors
• black permanent marker		• 2 clear containers with lids	• tape

Steps:
1. Make two loops of blue yarn by tying the ends of each of two strands together. These represent bacterial chromosomes.
2. Make a loop of green yarn by tying the ends of a strand together. This represents a plasmid.
3. Label one of the plastic containers *Bacterium A*. Place one bacterial chromosome and the plasmid into this container.
4. Label the second plastic container *Bacterium B*. Place one bacterial chromosome into this container.
5. Cut a piece of red yarn to represent human DNA. Label a section of this yarn *Human Insulin Gene*.
6. Using the marker, make black dots on either side of the label.
7. Label the scissors *Enzyme A*.
8. Label the tape *Enzyme B*.
9. Extract the plasmid from the Bacterium A container.
10. Use Enzyme A (the scissors) to cut the plasmid in one place. Cut the human DNA (red yarn) at both ends of the Human Insulin Gene (black dots).
11. Insert the Human Insulin Gene into the plasmid. Secure it in place with Enzyme B (the tape).
12. Place the recombinant DNA into Bacterium B.

ANALYZE AND CONCLUDE:

1. What is a plasmid? _____

_____

2. What is recombinant DNA? _____

_____

3. What object from this activity represented recombinant DNA? _____

_____

Lab 6.2.1A Manufacturing Insulin

4. How is recombinant DNA constructed? _____

_____

_____

5. Draw a picture of how the recombinant DNA was constructed. Label your diagram.

6. Why do you think that bacteria are used to copy the gene? _____

_____

7. Why must the human gene be inserted into a plasmid instead of directly into the

bacterial cell? _____

_____

8. Why might scientists want to produce plasmids with the human insulin gene?

_____

_____

Lab 6.2.2A Dolly, the Cloned Sheep

QUESTION: How was cloning used to create Dolly the sheep?

HYPOTHESIS: _____

_____

EXPERIMENT:

You will need:	• jelly bean (same color as	• safety pins	• small plastic container
• paper bag	plastic egg)	• rubber bands	with lid
• plastic egg	• paper clips	• forceps	• black jelly bean

Steps:

1. Your teacher will have the necessary items prepared ahead of time. Open the paper bag, which represents a uterus, and remove the plastic egg, which depicts an egg cell.
2. Carefully open the egg cell. The jelly bean inside represents the nucleus. The cell's cytoplasm also includes mitochondria (paper clips), ribosomes (safety pins), and the Golgi apparatus (rubber bands).
3. Use the forceps to remove the nucleus from the egg cell without disturbing the cytoplasm.
4. Carefully open the small plastic container labeled *Somatic Cell*. Use the forceps to remove the nucleus (black jelly bean).
5. Place the nucleus of the somatic cell into the egg cell. Ask your teacher to supply the "electric pulse" needed by the cell so the nucleus fuses with the egg cell and the new cell begins to divide. Close the egg.

ANALYZE AND CONCLUDE:

1. What type of cell was the donor cell? _____

2. What type of cell was the recipient cell? _____

3. Why do you think the scientists who cloned Dolly chose that type of cell to be the recipient cell? _____

_____

4. Why was the electric pulse needed? _____

_____

5. The somatic cell from which Dolly was cloned came from an adult ewe (sheep). Do you think Dolly's cells were all adult cells even though she was a newborn sheep? Why? _____

Lab 6.2.2A Dolly, the Cloned Sheep

Research to find the following information:

6. Why was Dolly the sheep unique? _____

_____

7. How many attempts were made before scientists had success with cloning Dolly?

_____

8. Knowing the number of attempts before success, how can you apply that to your

own life? _____

_____

9. After the egg cell was cultured for several days and it became an embryo, what

did scientists do next? _____

_____

10. Sheep can live about 11 to 12 years. Why was Dolly's life half as long? _____

_____

11. Where did the research and cloning process to create Dolly take place? _____

_____

Genetics Crossword

Genetics Crossword continued

Across

5. An agent used to get recombinant DNA into an organism's nucleus is a _____.

7. A group of cells that is genetically identical is a _____.

8. Genetic engineering can involve using viruses that attack bacteria. These viruses are called *bacteriophages* or _____.

9. A living thing (such as corn) whose DNA has been changed by adding or taking away genes is a _____ _____ _____.

10. The process of changing a cell's genes is _____.

11. The mating of closely related organisms, such as dogs from the same litter, is _____.

Down

1. The intentional crossing of strawberries with large, sweet berries to produce offspring with these traits is an example of _____ _____.

2. Replacing a nucleus and its DNA from one cell with the nucleus from another cell is _____ _____.

3. A DNA molecule formed from pieces of DNA from more than one species (usually in order to combine beneficial characteristics of two species) is _____ _____.

4. The point along the DNA at which one nucleotide pair is different from that of another person's DNA is called *a single nucleotide* _____.

6. Genetic engineering can involve using rings, or loops, of bacterial DNA called _____.

9. The process in which genes are transferred from one organism to another is called _____ _____.

Lab 7.1.4A Freshwater Organisms

QUESTION: How does pond life compare with stream life?

HYPOTHESIS: _____

_____

EXPERIMENT:

You will need:	• stream water	• thermometer	• 2 microscope slides	• microscope
• pond water sample	sample	• eyedropper	• 2 coverslips	• field guide

Steps:

1. Collect water samples from a pond and a stream. Record the water temperature of each sample. Pond _____ Stream _____

2. Describe the water in each sample.

   Pond _____

   Stream _____

3. What type of organisms do you expect to find in the pond water sample?

   _____

4. What type of organisms do you expect to find in the stream sample?

   _____

5. Use an eyedropper to place some pond water on a microscope slide. Place a coverslip over the pond water. Using a microscope, search the sample for rotifers, hydra, flatworms, segmented worms, crustaceans, insect larvae, sponges, protozoa, and algae.

6. Use a field guide or another scientific resource to identify the organisms you find. Record these organisms in the data chart on **WS 7.1.4A Freshwater**.

7. Rinse your eyedropper. Use the same procedure to search the stream water sample for organisms. Refer to a field guide to identify the organisms. Record the organisms on the data chart.

ANALYZE AND CONCLUDE:

1. Did you find the same types of organisms in the pond water as in the stream

water? _____

How did the number of organisms compare? _____

_____

Lab 7.1.4A Freshwater Organisms

2. What similarities did you observe? _____

_____

3. What accounts for these similarities? _____

_____

4. What differences did you observe? _____

_____

5. What might account for these differences? _____

_____

6. Compare your predictions in Steps 3–4 to what you actually observed.

_____

_____

7. Compare your original hypothesis to what you observed. _____

_____

8. Rewrite your hypothesis (if necessary) using what you have learned from this

experiment. _____

_____

_____

Name: _____   Date: _____

Lab 7.1.5A Stream Characteristics

QUESTION: What are the physical and chemical characteristics of a stream?

HYPOTHESIS: _____

_____

EXPERIMENT:

You will need:	• stopwatch	• dissolved oxygen	• pH test kit	• thermometer
• 50 m tape measure	• fishing bobber	test kit	• water hardness kit	
• meterstick	• 6 test tubes	• ammonia test kit	• nitrate test kit	

Steps:

1. Record your observations on the Stream Characteristics chart on the next page as you complete the steps below.
2. Use the meterstick to measure the depth of the stream's center. Record the depth.
3. Use the 50 m tape measure to measure the width of the stream. Stand on one side of the stream holding the tape. Have your partner hold the other end of the tape and walk across to the bank on the other side of the stream. Record the width. If this is not possible, find a place nearby that is about the average width of the stream. Measure the width in meters, including decimal points.
4. Mark off a 10 m distance along the bank of the stream. Be sure the distance is a straight line. With your stopwatch, time how long it takes for the fishing bobber to travel 10 m. Divide this number by 10 to determine how many seconds it took the bobber to travel 1 m. From this information, do the proper conversions to determine kph. Record this figure in the data chart.
5. Fill a test tube with stream water. Hold an object (such as a pen) behind the test tube, and use the scale below to determine a turbidity value. A turbidity value is a measure of water clarity. Record the value.
    1. clear with a few floating particles
    2. slightly cloudy; can still see the object on the other side of the test tube
    3. cloudy; object behind test tube is very blurry
    4. cannot see the object except a vague outline
6. Collect water in a clean test tube. Determine the dissolved oxygen (DO) level in ppms by following the directions on your test kit. Record the amount.
7. Collect another sample of water in a different test tube. Determine the ammonia (NH_3) level by following the directions on your test kit. Record the amount.
8. Using water in a fourth test tube, determine the pH level by following the directions on your test kit. Record the amount.
9. Fill another test tube with water and determine the water hardness level by following the directions on your test kit. Record the amount.
10. In the sixth test tube, collect water and determine the nitrate level by following the directions on your test kit. Record the amount.
11. Without touching the thermometer to anything but water, hold the thermometer beneath the surface of the water for 1 minute. Record the temperature.

Lab 7.1.5A Stream Characteristics

Stream Characteristics

Stream Depth (cm)	
Stream Width (m)	
Water Speed (kph)	
Turbidity (clarity)	
Dissolved Oxygen	
Ammonia	
pH Level	
Hardness	
Nitrate Level	
Temperature	

ANALYZE AND CONCLUDE:

1. How is the temperature of freshwater related to its dissolved oxygen? Draw a line graph illustrating this relationship. Label the axes and add a title.

_____

_____

_____

_____

_____

_____

2. What could cause the temperature of stream water to rise and how would this change the ecosystem? _____

_____

3. What could make the stream flow faster or slower? _____

_____

4. What could make the turbidity value change? _____

_____

_____

Lab 7.1.5B Pollution Test

QUESTION: How can I tell if the water is polluted?

HYPOTHESIS: _____

_____

EXPERIMENT:

You will need:	• ammonia test kit	• water hardness test kit
• 3 water samples from various sources	• nitrate test kit	

Steps:
1. One the data chart, list the samples and their predicted and actual levels.

Sample	Predicted Ammonia Level	Actual Ammonia Level	Predicted Nitrate Level	Actual Nitrate Level

2. Predict which samples will have the highest level of ammonia and nitrates. Record your predictions.
3. Follow the directions on the chemical test kits to determine the pollution levels of your three water samples and record the results on the data chart.
4. Determine the water hardness of your three samples. Record the results on the chart.

Sample	Hardness

ANALYZE AND CONCLUDE:

1. What might happen to fish if there was too much ammonia in their water? _____

_____

2. Use the information in *FYI: Effects of Ammonia on Fish* in your textbook to analyze

the effect each of your water samples would have on fish. _____

_____

3. What could happen to fish and people if there were too many nitrates in the water?

_____

_____

Lab 7.1.5B Pollution Test

4. Analyze the amounts of nitrates in your samples to determine whether any of them would be harmful to fish or people. _____

_____

5. Compare the hardness of the water samples. _____

_____

6. How might streams and lakes become more acidic? _____

_____

7. Of the three water samples you tested, which one would be the safest to drink?

_____

8. Of the three water samples you tested, which one would be the best water for fish to live in? _____

Habitat Search

Go outside to a park, field, or open space and see how many species of organisms you can find. Using a field guide, write the name of each species of organism in the data chart below. Record where you found each species, what its preferred habitat is like, and what basic needs are met in its habitat.

Species	Where It Was Found	Preferred Habitat	Basic Needs Met in Preferred Habitat

Population Density   **WS 7.1.2A**

1. Determine the area of your classroom in square meters. Divide the number of people in the room by the area of the room. What is the population density of the classroom? Show your work. _____

2. Find out the area of your school in square meters. Determine the population density of your school. _____

3. Research to find the area of the town or city you live in. Determine its population density. _____

4. Determine the population density for your state, province, or country. _____

5. Compare the population densities of your classroom, school, town or city, and state, province, or country. _____

   _____

   _____

6. Find the latest information on population densities by country. Compare the population density of your country to at least eight other countries (not listed in your textbook). _____

   _____

   _____

Freshwater

Use the data chart below to record your observations from **Lab 7.1.4A Freshwater Organisms**.

Pond and Stream Life Data Chart

Organism Name	Sketch	Stream or Pond

Name: _____   Date: _____

Stream Characteristics   **WS 7.1.5A**

Using the stream pictured, substitute the following for Steps 1–4 on **Lab 7.1.5A Stream Characteristics**.

Steps:

1. Record your observations on the chart on **Lab 7.1.5A Stream Characteristics**.

2. Measure and record the depth at the center of the depicted stream. _____

3. Measure and record the width of the depicted stream. _____

4. It takes a fishing bobber 22 seconds to travel 10 m on the stream. From this information, do the proper conversions to determine kph. Record this figure in the data chart. _____

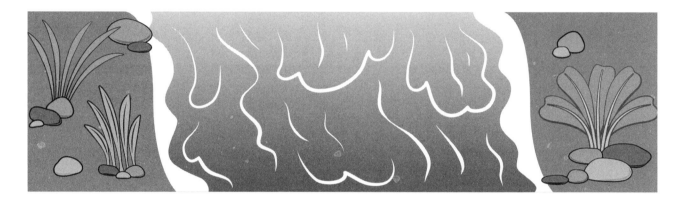

Scale: 1 cm = 1.2 m

Review

1. Name three abiotic factors in a stream. _____

_____

_____

2. Name three biotic factors in a stream. _____

_____

_____

3. Name three freshwater ecosystems in your region. _____

_____

_____

4. Compare the members of a community with the members of a population. _____

_____

_____

5. Choose one organism from each kingdom and name its preferred habitat.

　　a) Animal _____

　　b) Plant _____

　　c) Fungus _____

　　d) Protist _____

6. Give three examples of pollutants common to your region. _____

_____

_____

7. What might cause the carrying capacity of deer in a forest to drop? What might

cause it to rise? _____

_____

_____

8. What is the population density of 200 lemmings living on 25 hectares of tundra?

_____

9. Explain why the pH of a stream in an unpolluted area can become more acidic.

_____

_____

_____

10. Describe a process that might cause sediments from a field to pollute a stream.

_____

_____

_____

11. Why are wetlands important? _____

_____

_____

12. What affects water hardness? _____

_____

_____

13. Give an example of how helping one part of the biosphere helps another part.

_____

_____

_____

Name: _____ Date: _____

Lab 7.2.1A Food Web Interactions

QUESTION: How do the organisms in a food web interact?

HYPOTHESIS: _____

_____

EXPERIMENT:

You will need:	
• hand lens	• field guides to plants, fungi, and insects

Steps:
1. Search an ecosystem for as many species of organisms as you can find. Look in forest leaf litter, in the cracks of tree bark, beneath rocks and leaves in a stream, under leaves on trees, beneath rotting logs, near the roots of tall grasses and weeds, and in the canopy or treetops of a forest.
2. Have one person in your group record all the organisms that you can readily identify. The other members of your group should use field guides to identify organisms with which you are not familiar. Remember to include all types of organisms, not just animals.
3. Describe the niche (including habitat and trophic level) of each organism.

Organism	Niche

Lab 7.2.1A Food Web Interactions

4. Sketch as many food chains as you can to include the organisms you found.

ANALYZE AND CONCLUDE:

1. Which role do most of the organisms fill? _____

_____

2. Were you able to record all the organisms that live in that ecosystem? Why? _____

_____

_____

_____

_____

3. Do top predators exist in your ecosystem? If not, why? _____

_____

_____

_____

4. Describe how the organisms interact in your ecosystem. _____

_____

_____

_____

_____

Lab 7.2.1B Biomass

QUESTION: How can biomass be determined?

HYPOTHESIS: _____

_____

EXPERIMENT:

You will need:	• wide shovel	• fine mesh	• triple beam balance	• wide-mouthed jar
• meterstick	• 5-gallon bucket	• rubber band	• tarp or sheet	

Steps:

1. Determine and record the combined mass of the wide-mouthed jar, rubber band, and mesh. _____
2. Measure a square meter in a forest ecosystem. Use the shovel to lift 5–8 cm of leaf litter from the ground and place it into the bucket.
3. Place any insects you find into the wide-mouthed jar. Cover the jar with the fine mesh and secure it with a rubber band. Bring the insects and the leaf litter back to the classroom.
4. Weigh and record the mass of the insect-filled jar. _____

   Calculate the mass of just the insects. _____ Release the insects back into the forest.
5. Lay out the leaf litter on a sheet or tarp for two or three days to dry. Determine the mass of the dry leaf litter. _____
6. Add the mass of the dry leaf litter to the mass of the invertebrates you collected to estimate the biomass of the leaf litter in your square meter of forest.

   _____
7. Return the leaf litter to the forest.

ANALYZE AND CONCLUDE:

1. Was the process for determining biomass close to what you hypothesized? Why?

_____

_____

2. If you were to take the biomass of the leaf litter and multiply it by the total area of the forest, would you be determining the total biomass of the forest? Explain.

_____

_____

3. Why do you think determining biomass is important to ecologists? _____

_____

_____

Name: _____ Date: _____

Lab 7.2.4A Recolonization

QUESTION: How do living things recolonize a community?

HYPOTHESIS: _____

_____

EXPERIMENT:

| **You will need:** | • 4 tent stakes | • garden trowel |
| • 4.5 m string | • meterstick | • field guide |

Steps:
1. Using string and 4 tent stakes, rope off a square meter patch of leaf litter in a wooded area.
2. Use a garden trowel to collect and categorize every organism you find within the leaf litter inside your square. Search at least 10 cm down into the litter. Remember that organisms include plants and fungi.
3. Record data on the organisms you find.

Organism	**Description**	**Recolonize the Square?**

Lab 7.2.4A Recolonization

4. Release any animals at least 6 m from the square.
5. Replant the small plants and fungi at least 6 m away from the square.
6. Revisit the square every other day, recording what organisms recolonize the square.
7. After 10–14 days of observations, compare the different species and number of individuals that recolonized your square with the original diversity you found there. _____

_____

_____

_____

_____

ANALYZE AND CONCLUDE:

1. Which organisms (fungi, plants, or invertebrates) recolonized the square the fastest? _____

_____

_____

2. Which organisms were slowest to recolonize the square? _____

_____

_____

_____

3. Use **WS 7.2.4A Bar Graphs** to make bar graphs to determine how the composition of the organisms in the square changed.

4. What primary methods did the organisms use to recolonize the square? _____

_____

_____

_____

5. Was the recolonization in a primary or secondary succession area? _____

_____

_____

_____

Lab 7.2.4B Effect of Fertilizer

QUESTION: How does fertilizer affect an aquatic (pond) ecosystem?

HYPOTHESIS: _____

_____

EXPERIMENT:

You will need:	• fertilizer	• marking pencil	• 3 microscope slides
• 3 wide-mouthed jars	• 9 water plants, such as	• teaspoon	• 3 coverslips
• pond water	elodea	• microscope	

Steps:
1. Fill each jar with pond water. Label the jars as *Jar 1*, *Jar 2*, and *Jar 3*.
2. Place three plants in each jar.
3. Add ½ tsp of fertilizer to Jar 1 and 4 tsp to Jar 2. Do not add any fertilizer to Jar 3.
4. Place the jars in a well-lighted place.
5. Predict what will happen to the plants in each jar. Record your predictions.

Jar 1 _____

_____

Jar 2 _____

_____

Jar 3 _____

_____

Observe the jars daily for two weeks. Record your observations.

Week 1	Jar 1	Jar 2	Jar 3
Day 1			
Day 2			
Day 3			
Day 4			
Day 5			

Lab 7.2.4B Effect of Fertilizer

Week 2	Jar 1	Jar 2	Jar 3
Day 1			
Day 2			
Day 3			
Day 4			
Day 5			

6. Examine the water from each jar under a microscope. Record your observations.

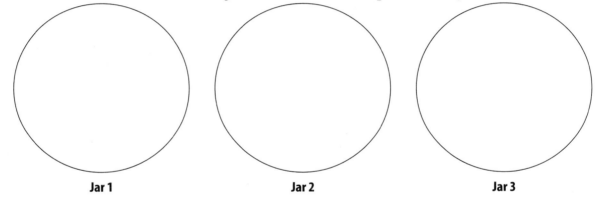

Jar 1   **Jar 2**   **Jar 3**

ANALYZE AND CONCLUDE:

1. At the end of the experiment, what were the differences in the plants in each jar?

_____

_____

2. Too much algae in the water can increase eutrophication, a natural process that decreases the amount of dissolved oxygen in the water. Explain how fertilizer might affect the aquatic ecosystem as a whole. _____

_____

_____

3. Compare your hypothesis to the actual results. _____

_____

_____

Trophic Level

Search an ecosystem near your school or neighborhood and identify the organisms you find by their trophic level. Woodlands and meadows will provide the best results. Use the chart below to record and identify producers, primary consumers (herbivores), secondary and tertiary consumers (carnivores), and detritivores (scavengers and decomposers). Use a field guide and a hand lens to find and identify smaller organisms.

Producers	Primary Consumers	Secondary Consumers	Tertiary Consumers	Detritivores

Food Chain

Make a list of all the foods you ate yesterday. Determine each food's place in the food chain. List the producers and the consumers. If possible, determine whether the consumers are primary, secondary, or tertiary consumers.

Food Eaten	Place in the Food Chain

Bar Graphs

Make bar graphs to determine how the composition of the organisms in the square selected for **Lab 7.2.4A Recolonization** changed. Make a different graph for each species of organism, using a red bar to represent the number of individuals from that species found in the square before their removal and a green bar to represent the number of individuals found in the square after their removal. Label each graph with the type of organism and label the axes.

Review

Indicate whether each statement is true or false. Explain why false statements are false.

1. An ecosystem's biomass is the total weight of its plants. _____
_____

2. A ladybug eats a mealworm and an aphid, so the ladybug is a carnivore. _____
_____

3. Because it eats insects, a Venus flytrap is a consumer as well as a producer. _____
_____

4. Decomposers break down live tissues of plants, animals, and fungi. _____
_____

5. Detritivores such as snails eat flesh. _____
_____

6. A hawk eats a snake, which has eaten a toad and a mouse. The toad ate a cricket, a fly, and an earthworm. The mouse ate seeds and grain. The crickets ate a rotten tomato and insects. The earthworm ate manure and rotting leaves in the soil. This is an example of a food chain. _____
_____

7. A horse, which eats only plants, is an omnivore. _____
_____

8. A cow eats grass, so the cow is a primary consumer. _____
_____

9. A weasel eats a rabbit, which ate lettuce. The weasel is a secondary carnivore.
_____

10. The spotted owl, which has no predators in a forest, is a top carnivore. _____
_____

11. Many plants and animals can share one niche. _____
_____

12. A mushroom that absorbs its food from a rotting log is a producer. _____
_____

13. Only one organism occupies each trophic level. _____
_____

14. Mutualism only occurs among animals. _____
_____

15. Many desert plants have very long roots that reach down to the water table, allowing the plant to get enough water. This is an example of symbiosis. _____
_____

16. Many butterflies look a lot like monarch butterflies, which are poisonous. This is a moment-of-truth defense. _____
_____

17. Animals such as bears and chipmunks enter a period of inactivity during the winter. This state is called *hibernation*. _____
_____

18. The process of gradual change in an ecosystem—such as sunlight reaching the forest floor because trees have fallen, which allows raspberries to grow—is ecological succession. _____
_____

19. The growth of plants on land void of plants is primary succession. _____
_____

20. When a spider catches a fly in its web and eats it, the spider is the prey and the fly is the predator. _____
_____

Lab 7.3.1A Biomes and Seed Growth

QUESTION: How do different biomes affect seed growth?

HYPOTHESIS: _____

_____

EXPERIMENT:

You will need:	• native desert seeds (such as chia, desert paintbrush, or desert chicory)	• native deciduous forest seeds (such as maple, oak, or violets)	• native grassland seeds (such as Indian grass, buffalo grass, or rye)
• cardboard box			
• scissors			• metric ruler
• soil or sand	• native coniferous forest seeds (such as redwood, sorrel, or wintergreen)	• native rain forest seeds (such as gardenia flower or devil's chili)	• lamp
• salt			• cookie sheet
• pine needles			• tape
• leaf litter	• water	• plastic wrap	

Steps:
1. Use the scissors to carefully poke several holes in the bottom of the box.
2. Fill the box with at least 10 cm of soil.
   • If you are building a desert biome, fill the box with a mixture of one-fourth sand and three-fourths soil. Sprinkle it lightly with salt and mix the salt with the soil.
   • If you are building a coniferous forest biome, spread a thin layer of pine needles on top of the soil, and mix them in. After you plant your seeds, spread another thin layer of pine needles over the top.
   • If you are building a deciduous forest biome, mix leaf litter with the soil.
3. Divide the box into five sections. Label the sections *Coniferous Forest Seeds*, *Deciduous Forest Seeds*, *Rain Forest Seeds*, *Grassland Seeds*, and *Desert Seeds*.
4. Plant appropriate seeds in each section. Record how many seeds of each kind you have planted.
5. Place the box on a cookie sheet. Water the seeds, and cover the top of the box with plastic wrap. Tape the plastic wrap to the box.
6. Keep the box in a warm place in which you can use the lamp to control the light.
7. For 14 days, record the seed growth for each kind of seed on the data chart on **WS 7.3.1A Seed Growth**.
8. Predict which kind of seeds will grow best in your biome. Record your predictions.
9. Predict how the seeds will grow in the other biomes built by other groups. Record your predictions.
10. When the seeds have sprouted, follow the instructions for your biome.
    • Coniferous forest: Let the surface dry and then add water. Use the lamp to provide 1–2 hours of light per day.
    • Deciduous forest: Let the surface dry, and then add water. Use the lamp to provide 2–3 hours of light per day.

Lab 7.3.1A Biomes and Seed Growth

- Rain forest: Keep the surface wet at all times. Do not give the box direct light.
- Grassland: Let the surface dry, and then add water. Use the lamp to provide 5–6 hours of light per day.
- Desert: Let the soil dry to a depth of 2.5 cm. Use the lamp to provide 6–8 hours of light per day.

ANALYZE AND CONCLUDE:

1. Look at the different biomes. Describe the growth of the different seeds in each biome model. _____

_____

_____

2. In which biome did each kind of seed grow best? _____

_____

3. Which seeds grew well in more than one biome? _____

_____

_____

4. How did each seed react to the allotted amount of light? _____

_____

_____

5. What factors besides light affected seed growth? _____

_____

_____

6. In what ways were the biome models like real biomes? _____

_____

_____

7. What important factors of real biomes were not included in the biome models?

_____

_____

8. How could you design a tundra biome? _____

_____

_____

Seed Growth

Use the information on the data chart from **Lab 7.3.1A Biomes and Seed Growth** to record your observations of seed growth.

Record the type of biome your group created. _____

	Coniferous Seeds	Deciduous Seeds	Rain Forest Seeds	Grassland Seeds	Desert Seeds
Day 1					
Day 2					
Day 3					
Day 4					
Day 5					
Day 6					
Day 7					
Day 8					
Day 9					
Day 10					
Day 11					
Day 12					
Day 13					
Day 14					

Alaska Survival

You and your friends are traveling along the Yukon River in the autumn on an expedition to observe caribou in central Alaska when your Arctic rover runs out of gas. You are 100 km from your research station. Your radio is frozen and your homing beacon is broken. You have no way of communicating with your fellow scientists. You and your friends must travel across the snow-covered tundra to reach your home base. You can only bring 10 items with you.

Choose the 10 items from the equipment list that you think your group will need to cross the frozen, snow-covered tundra. List the items in order of importance. Next to each item briefly explain why you chose that item.

Item	Reason for Choosing It
1.	
2.	
3.	
4.	
5.	
6.	
7.	
8.	
9.	
10.	

1.	fur parkas with hood	**21.**	insect repellent
2.	heavy wool and fleece clothing	**22.**	malaria pills
3.	lightweight cotton clothing	**23.**	waterproof container of matches
4.	lightweight cotton hats	**24.**	steel wool
5.	rain gear	**25.**	10 comic books
6.	sunglasses	**26.**	bundle of wood in plastic trash bag
7.	fur-lined boots	**27.**	multi-purpose pocketknife
8.	hiking boots	**28.**	snake bite kit
9.	2 nylon tents	**29.**	hand gun
10.	ice block maker	**30.**	20 bullets
11.	down sleeping bags	**31.**	flare gun
12.	lightweight sleeping bags	**32.**	2 flares
13.	extra blankets	**33.**	compass
14.	cross-country skis	**34.**	flashlight and batteries
15.	snowshoes	**35.**	shovel
16.	rubber raft built for six people	**36.**	mirror
17.	five days' supply of food	**37.**	salt pills
18.	12 canteens of water	**38.**	first-aid kit
19.	fishing lines and bait	**39.**	cooking stove and fuel
20.	water purifying pills	**40.**	machete

Name: _____   Date: _____

Amazon Survival

You and your friends are flying over the Amazon rain forest in Brazil on a special zoological expedition when your plane malfunctions and you have to make an emergency landing in the jungle. Looking at a map of the region, you discover you are 1 km from the Rio Roosevelt, a branch of the Amazon River. You are 240 km from the nearest village. Your radio is damaged, and no one knows where you are. You and your friends must try and cross 240 km of rain forest to get help.

Choose the 10 items from the equipment list that you think your group will need to survive the Amazon rain forest. List the items in order of importance. Next to each item explain why you chose that item and why you placed it where you did on the list.

Item	Reason for Choosing It
1.	
2.	
3.	
4.	
5.	
6.	
7.	
8.	
9.	
10.	

1.	fur parkas with hood		**21.**	insect repellent
2.	heavy wool and fleece clothing		**22.**	malaria pills
3.	lightweight cotton clothing		**23.**	waterproof container of matches
4.	lightweight cotton hats		**24.**	steel wool
5.	rain gear		**25.**	10 comic books
6.	sunglasses		**26.**	bundle of wood in plastic trash bag
7.	fur-lined boots		**27.**	multi-purpose pocketknife
8.	hiking boots		**28.**	snake bite kit
9.	2 nylon tents		**29.**	hand gun
10.	ice block maker		**30.**	20 bullets
11.	down sleeping bags		**31.**	flare gun
12.	lightweight sleeping bags		**32.**	2 flares
13.	extra blankets		**33.**	compass
14.	cross-country skis		**34.**	flashlight and batteries
15.	snowshoes		**35.**	shovel
16.	rubber raft built for six people		**36.**	mirror
17.	five days' supply of food		**37.**	salt pills
18.	12 canteens of water		**38.**	first-aid kit
19.	fishing lines and bait		**39.**	cooking stove and fuel
20.	water purifying pills		**40.**	machete

Desert Survival

You and your friends are flying from Morocco to Egypt when you crash-land somewhere in the country of Libya in the middle of a desert. You are 150 km from the nearest village. Your radio is broken beyond repair and no one knows where you are. You and your friends must try to cross 150 km of Libyan desert to get help.

Choose 10 items from the equipment list you think your group will need to survive in the desert. List the items in order of importance. Next to each item explain why you chose that item and why you placed it where you did on the list.

Item	Reason for Choosing It
1.	
2.	
3.	
4.	
5.	
6.	
7.	
8.	
9.	
10.	

#	Item	#	Item
1.	fur parkas with hood	21.	insect repellent
2.	heavy wool and fleece clothing	22.	malaria pills
3.	lightweight cotton clothing	23.	waterproof container of matches
4.	lightweight cotton hats	24.	steel wool
5.	rain gear	25.	10 comic books
6.	sunglasses	26.	bundle of wood in plastic trash bag
7.	fur-lined boots	27.	multi-purpose pocketknife
8.	hiking boots	28.	snake bite kit
9.	2 nylon tents	29.	hand gun
10.	ice block maker	30.	20 bullets
11.	down sleeping bags	31.	flare gun
12.	lightweight sleeping bags	32.	2 flares
13.	extra blankets	33.	compass
14.	cross-country skis	34.	flashlight and batteries
15.	snowshoes	35.	shovel
16.	rubber raft built for six people	36.	mirror
17.	five days' supply of food	37.	salt pills
18.	12 canteens of water	38.	first-aid kit
19.	fishing lines and bait	39.	cooking stove and fuel
20.	water purifying pills	40.	machete

Biomes

Write the correct term in the blank.

1. A region such as a desert, grassland, or deciduous forest is called a

_____.

2. A _____ is a condition that limits the number of organisms in an area.

3. _____ is the layer of soil in the tundra that never melts.

4. Trees that have needles and produce cones are _____.

5. Trees that drop their leaves at the end of the growing season are known as

_____.

6. Earthworms and other organisms that break up leaf litter by passing it through their bodies are called _____.

7. _____ is the dead and decaying organic matter that covers the soil.

8. The number and variety of plants, animals, fungi, and other organisms in an area is called _____.

9. Spanish moss and other plants that live in the branches of trees are called

_____.

10. Savannas, pampas, prairies, and steppes are considered _____.

11. _____ is a process that creates new deserts through climate change or destructive land use.

12. _____ is the visible light given off by fireflies and many ocean organisms.

13. The area where the tides go in and out is called an _____.

14. The _____ is the area over the continental shelf.

15. Ocean plants can photosynthesize in the _____.

16. _____ are microscopic aquatic plants.

17. Microscopic aquatic animals are called _____.

Extinction

Use **BLM 7.3.9A Extinction Threats** to answer the following questions:

1. Which of these organisms did God declare to be good and command to be fruitful and multiply? _____

_____

2. The animals on the list are in danger for many reasons. Which of these reasons can people easily control without disrupting their lifestyle or livelihood? _____

_____

3. Which of the reasons for endangerment would require people to change their lifestyle? How could people do this? _____

_____

_____

4. What were the three most common reasons for species endangerment?

_____

_____

_____

5. Do you think it is always practical for people to make sacrifices to save a species?

_____

_____

_____

_____

6. If people cannot or do not want to make these sacrifices, what could they do instead to help save the animals or plants? _____

_____

_____

7. Which of the reasons for species endangerment can people not control?

_____

_____

_____